SpringerBriefs in Applied Sciences and Technology

SpringerBriefs present concise summaries of cutting-edge research and practical applications across a wide spectrum of fields. Featuring compact volumes of 50 to 125 pages, the series covers a range of content from professional to academic.

Typical publications can be:

- A timely report of state-of-the art methods
- An introduction to or a manual for the application of mathematical or computer techniques
- A bridge between new research results, as published in journal articles
- A snapshot of a hot or emerging topic
- An in-depth case study
- A presentation of core concepts that students must understand in order to make independent contributions

SpringerBriefs are characterized by fast, global electronic dissemination, standard publishing contracts, standardized manuscript preparation and formatting guidelines, and expedited production schedules.

On the one hand, **SpringerBriefs in Applied Sciences and Technology** are devoted to the publication of fundamentals and applications within the different classical engineering disciplines as well as in interdisciplinary fields that recently emerged between these areas. On the other hand, as the boundary separating fundamental research and applied technology is more and more dissolving, this series is particularly open to trans-disciplinary topics between fundamental science and engineering.

Indexed by EI-Compendex, SCOPUS and Springerlink.

More information about this series at http://www.springer.com/series/8884

Vincent Kvočák · Daniel Dubecký

Research and Development of Deck Bridges

 Springer

Vincent Kvočák
Faculty of Civil Engineering
Technical University of Košice
Košice, Slovakia

Daniel Dubecký
Faculty of Civil Engineering
Technical University of Košice
Košice, Slovakia

ISSN 2191-530X   ISSN 2191-5318 (electronic)
SpringerBriefs in Applied Sciences and Technology
ISBN 978-3-030-66924-9   ISBN 978-3-030-66925-6 (eBook)
https://doi.org/10.1007/978-3-030-66925-6

This Springer imprint is published by the registered company Springer Nature Switzerland AG
The registered company address is: Gewerbestrasse 11, 6330 Cham, Switzerland

Preface

The book is focused on deck bridges with encased steel beams. The introductory chapters of this book discuss the design process in deck bridges in the past, and present a selection of current issues regarding the design and construction of this type of bridge in Slovakia today. The theoretical part reflects the latest achievements of international endeavours in composite bridge research. The following experimental part provides new knowledge in the field obtained by research into structures with encased steel beams. This section of the book is exclusive in the sense of being entirely elaborated and based upon vast experiments carried out by the Faculty of Civil Engineering at the Technical University of Kosice. Fifty composite beams with rigid steel reinforcement of various sections under five different types of composite action have been experimentally investigated. The book further describes one specific type of composite action, namely a steel box section perforated in both its web and top flange. A series of static, long-term and fatigue tests have been conducted, and the outcomes are presented in the experimental section of the book. The results obtained are then compared with numerical simulations on selected models and analytical calculations complying with the Eurocode. This book also contains some information on testing the material properties of steel and concrete and their characteristics. Finally, a variety of types of composite action between steel and concrete have been examined, using push-out tests—and the final chapters summarise, examine and compare the results obtained.

The research findings presented have been achieved during the ITMS project No. 26220220124, "The Development of Bridges with Encased Filler-Beams of Modified Sections".

This book is intended for professionals working in the field of design and construction of composite steel and concrete bridges and also for scientists, researchers, designers and students who specialise in the field of bridge structures.

Košice, Slovakia

Vincent Kvočák
Daniel Dubecký

Funding Information

The book submitted has been supported by The projects: VEGA 1/0172/20 "Stress and deformation analysis of load bearing components made out of steel, glass and composite materials" of the Scientific Grant Agency of the Ministry of Education, science, research and sport of the Slovak Republic and the Slovak Academy of Sciences and by the project Slovak Research and Development Agency under the contract No. APVV-15-0486 "Analysis of Shear Connection Influence in Bridges with Encased Beams"

Introduction

Deck bridges with encased steel beams have been applied in bridge construction for many decades, while I-sections are even today predominantly used in steel beams as rigid load-bearing reinforcement. However, their use in the compression zone of the beam appears uneconomical. Issues regarding composite structures are nowadays increasingly debated. All building materials have some advantages and disadvantages; therefore, their right combinations, when selected and designed well, can eliminate the harmful properties of one specific type and employ the desirable properties of another. This book seeks to cast a light on the composite action of the two most frequently used materials—steel and concrete, in particular, a steel filler-beam encased in a concrete deck, the kind of which provides the most efficient outcome.

The book provides some information on the history of design and construction of composite members made of steel and concrete and familiarises its readers with current practical applications of such composite structures. Some valuable information on the previous and current methods and procedures in the design of deck bridges is also presented in the book, bringing some examples of real constructions from Slovakia and the rest of the world. Part of the work is devoted to experimental analysis, where the underlying theoretical assumptions and calculations are experimentally verified, proving our approach correct. Several partial results include the design of working models in a sophisticated software program that can precisely analyse the individual models.

Contents

Symbols and Abbreviations

A_c	Cross-sectional area of concrete
A_a	Cross-sectional area of the structural steel section
A_f	Cross-sectional area of the flange
A_w	Cross-sectional area of the web
E_a	Modulus of elasticity of structural steel
E_c	Modulus of elasticity of concrete
EI_i	Flexural stiffness of an ideal cross-section
F	Maximum force at the failure of test samples
$F_{pl,\,Rk}$	Maximum force at the failure of test specimens
$F_{max,\,exp}$	Measured values of the maximum forces
F_{exp}	Measured values of forces
I_a	Second moment of area of the structural steel section
$M_{max,\,exp}$	Experimentally determined values of maximum resistance moments
$M_{max,\,exp,\,ave}$	Average values of the experimentally determined maximum resistance moments
M_{exp}	Experimentally determined values of resistance moments
$M_{exp,\,ave}$	Average values of the experimentally determined resistance moments
$M_{pl,\,Rk}$	Characteristic value of the plastic resistance moment of the composite section
$M_{pl,\,Rd}$	Design value of the plastic resistance moment of the composite section
N_f	Number of concrete studs
P	Resistance of a concrete stud
S_a	Surface area of the steel beam
d_1, d_2	Dimensions of test moulds
h, b, d	Dimensions (height, width and depth) of test specimens
h_k	Height of a concrete stud
f_c	Characteristic value of the cube compressive strength of concrete
f_{cd}	Design value of the cylinder compressive strength of concrete
f_{cf}	Characteristic value of the flexural tensile strength of concrete
f_{ck}	Characteristic value of the cylinder compressive strength of concrete at 28 days
f_{ctk}	Characteristic value of the axial tensile strength of concrete

f_y	Nominal value of the yield strength of structural steel
f_{yd}	Design value of the yield strength of structural steel
f_{yf}	Nominal value of the yield strength of the structural steel of flanges
f_{yw}	Nominal value of the yield strength of the structural steel of webs
g	Dead load
l	Span between the specimens
l_0	Distance between supporting cylinders
n_0	Modular ratio for short-term loading
q_a	Unit weight of structural steel
q_c	Unit weight of concrete
t	Thickness of the steel sheet
t_f	Thickness of a flange of the structural steel section
x	Position of the force from the nearest support
z_a	Position of the centre of gravity of the structural steel section
z_f	Position of the centre of gravity of the flange of the structural steel section
z_i	Position of the centre of gravity of the ideal cross-section
z_{pl}	Position of the plastic neutral axis of the structural steel section
z_w	Position of the centre of gravity of the web of the structural steel section
γ_a	Partial factor for structural steel
γ_c	Partial factor for concrete
δ	Deflection

Chapter 1
The Current Situation in Bridge Construction

Bridges are regarded as the top of civil engineering mastery. The latest knowledge and trends are applied in bridge design and construction so that all technological and architectural requirements are met, while also taking the economy (the funds available to the employer to build the bridge) and the actual construction into account. Special attention must be paid to the technical assessment and reconstruction design of the existing bridges, where numerous factors need to be considered, such as traffic diversions, standard stipulations, material and technological aspects and possibilities at the time of their original design and construction. Today's construction market offers all different kinds of materials, technologies and possibilities of building various types, dimensions and shapes of bridges [1] (Fig. 1.1).

One of the best possibilities in bridge design is a combination of materials employing their best characteristics. Such structures include composite beam bridges, comprising of a steel beam and concrete deck, and combining favourable steel characteristics in tension (its high tensile strength), and consistent qualities of concrete (its resistance to compression forces). This composite action of both materials allows the creation of an advantageous structural system for both road and railway bridges. Nevertheless, the economy and the advantages of the simple construction of this popular type of bridge remain underestimated in bridge engineering.

1.1 Design Methods for Deck Bridges

Deck bridges with encased filler-beams are designed mainly for small spans of roads or railways. This type of structure has been used in our country for over 120 years without considerable changes. Nowadays, they are still frequently employed in railway reconstructions. The design methodology at the time of their original construction assumed static action of steel members only. In early designs, rails

© The Author(s), under exclusive license to Springer Nature Switzerland AG 2021
V. Kvočák and D. Dubecký, *Research and Development of Deck Bridges*,
SpringerBriefs in Applied Sciences and Technology,
https://doi.org/10.1007/978-3-030-66925-6_1

Fig. 1.1 Modern constructions of composite deck bridges. *Source* ÖBB Österreichische Bundes-bahnen, ArcelorMittal Company

were used and later also riveted, rolled or welded beams came into use. Concrete played only a stiffening role in those days [2] (Fig. 1.2).

The above structural system was quite widespread at the beginning of the twentieth century and applied in more than two-thirds of then constructed bridges. In the 1970s, a new structural disposition was introduced, where deck bridges with encased beams started to be regarded as reinforced steel structures. Concrete served a static load-bearing function, acting in compression. These structures were designed in accordance with the general rules given by permissible stresses method, and the technical standards such as ČSN 73 2089 *Směrnice pro navrhování spřažených ocelobetónových nosníků*, ČSN 73 6205 *Navrhování ocelových mostních konstrukcí* and ČSN 73 6206 *Navrhování betonových a železobetonových mostních konstrukcí* were applied. The State Railways had their own directives, structural designs and calculation templates [3, 4].

Today, the design of bridge structures is governed by the ultimate limit states method and the STN EN 1994-2 standard. This standard contains structural requirements and design procedures and describes the verification of a bridge deck with encased beams made of rolled or welded I-sections. The outer surface of the bottom flange is not encased; thus, it initially serves as permanent, leave-in-place shuttering, and when combined structurally with the hardened concrete, it acts as a reinforcement. Several directives and regulations on the design of railway bridges drawing from Eurocodes are currently in preparation [5, 6].

Fig. 1.2 Deck bridges with encased solid web girders. *Source* ArcelorMittal Company

1.2 Deck Bridges Built in the Slovak Republic

The construction and maintenance of railways in Slovakia are within the scope of activity of the Railways of the Slovak Republic, including the construction and maintenance of railway bridges. A well-tested and commonly employed construction type is the deck bridge with encased steel beams. This type is popular primarily due to its little headroom, simple construction, extensive good experience of its exploitation and easy maintenance.

Deck bridges with encased filler-beams are used for short and mid-spans. From the structural point of view, they are designed as simply-supported beams. Rolled steel sections are usually employed; welded sections are designed only when the use of rolled sections cannot provide for the maximum headroom/passage height. The depth of the beams ranges within 1/30–1/20 of the overall span. They are spaced at an axial distance of approximately 400 mm. The concrete cover over the steel beams is 60–300 mm. Load-bearing reinforcement bars are designed in the transverse direction. The deck is assumed to act transversely as a simply-supported beam with protruding ends, supported by rails and loaded by the vertical forces transferred by steel beams. Transverse reinforcement is provided towards/at the top surface only. Shear is supposed to be transmitted by steel beams. If their shear resistance is not sufficient, stirrups are designed for the region reaching as far as to 1/4–1/3 of the span from the support. The space between the bottom flanges is filled with steel sheets or cement-wood boards. The sheets are not genuinely appropriate in terms of fatigue effects as they create and present unsuitable details at the regions of their connection with the flanges [1, 7, 8].

Many railway bridges and culverts with encased steel beams have been constructed in Slovakia, and many of them are now about to reach their planned service life; therefore, they are being refurbished. The bridges that have been under reconstruction recently include, for instance, the railway bridge at the station of Báhoň, the one on the railway from Michaľany to Palota near the village of Veľaty, another on the railway between Hodonín and Holíč and many more [9].

1.3   The Present Situation in Deck Bridges in the World

Deck bridges with encased steel sections are quite widespread all around Europe but also in North America. Every country used to apply its own national design and construction standards for deck bridges in the past. Nowadays, each member state of the European Union has to observe Eurocode 4, a set of unified rules for civil engineering structures, whose Part 2 regulates the design of composite concrete and steel bridges [10].

The situation in the Czech Republic was previously the same as in the Slovak Republic. These days, it is not quite the same: although the legislators in the Czech Republic are currently also preparing new directives and regulations for the design of railway bridges, deck bridges are considered to act as frame structures. On the contrary, in Slovakia, these are in principle designed as simply-supported beams. In Belgium, pre-stressed bridge decks with encased U-sections are applied in refurbishments. The bridge deck is formed with two pre-stressed I-sections connected with concrete. This type of bridge deck comes from some older types that were pre-stressed in two phases: first, the steel beam was elevated, loaded by two concentrated forces, and cast in concrete. Seven days later, the beam was pre-stressed again and entirely placed in concrete [8].

1.3.1   Sete-Frontignan Bridge, France

Composite bridges with encased steel beams were initially developed for railway bridges; nevertheless, over the recent years, this type of bridge has found its applications in road bridges too. It provides a robust, resistant and straightforward structure that does not require any highly specialised craftsmanship. Due to the high load-bearing capacity of the bridge of this kind, there are always substantial reserves and a right margin in case that the use and operational conditions on the bridge have changed. The structure with encased steel beams is suitable for the exchange of the existing deck bridges, where the shallow depth of the deck makes it easier to adjust its geometrical dimensions to the existing supports. Moreover, the structure

Fig. 1.3 Laying the beams on the piers over the existing railway track. *Source* ArcelorMittal Company

also suits bridge decks with limited headway of the bridge or very small structural height, as well as the structures with heavy traffic flow under the bridge. Fast and straightforward construction without the necessity to use temporary structural support/falsework affects the traffic under the bridge as little as possible [2, 11] (Fig. 1.3).

The construction of the Sete-Frontignan composite steel and concrete bridge over the railway caused only a minimum short-term closure of rail traffic. The beams had been manufactured in a plant in advance and transported to the construction site in several segments that were then laid on concrete piers. The whole structure was made monolithic afterwards [12, 13] (Fig. 1.4).

1.3.2   Cyrnos Bridge, Senegal

Simple production technology and fast construction of deck bridges with encased beams make it possible to build this type of bridge even at sites where the construction of more complex bridges would be infeasible or very complicated. Bridge bearings allow for movement, hence the dilatation in one direction as is displayed in Fig.1.5.

Fig. 1.4 Sete-Frontignan Bridge, France. *Source* ArcelorMittal Company

Fig. 1.5 Detail of the bridge bearing. *Source* ArcelorMittal Company

Fig. 1.6 Detail of the bridge bearing. *Source* ArcelorMittal Company

The application of encased steel beams reduces the construction costs in wider bridges in comparison with an orthotropic bridge deck, which is replaced with a cheaper concrete deck, as well as dramatically decreases maintenance costs. From the structural point of view, the effects of shrinkage, creep and temperature change can be accounted for more faithfully and reliably in composite structures [13, 14] (Fig. 1.6).

1.4   Application of Various Beam Shapes and Arrangements in Bridges

Rolled steel sections are laid longitudinally on pre-arranged bridge bearings. It is possible to use permanent or self-acting shuttering, which is then placed on the bottom flanges of steel beams. Consequently, transverse reinforcement made of plain rebars and tied to the desired shape is added. A significant advantage of this bridge deck is its variability, which enables engineers to adjust steel beams to the route of transport on the bridge while using various curvatures, as shown in Fig. 1.7 [2].

Furthermore, encased steel beams can also be employed in multi-span bridges, where the very structure of the bridge must be extremely rigid and also where stringent requirements are placed on the fatigue resistance of the material used in a bridge. One such multi-span deck bridge with encased steel beams was built for a high-speed railway over the River Moselle in France (Fig. 1.8).

Fig. 1.7 Curved steel beams, France. *Source* ArcelorMittal Company

Fig. 1.8 A multi-span railway bridge. *Source* ArcelorMittal Company

The structure is made of longitudinal steel beams laid on bridge bearings and then on cast-in-place concrete piers. The mass of the structure reliably resists the dynamic impacts produced during the passage of high-speed trains on the bridge [2, 14].

1.5 New Technology Used on Composite Bridge

The VFT-WIB ("Verbund-Fertigteil-Träger–Walzträger im Beton") technology was first used for the road bridge in Bad Vigaun, Austria. The construction was completed in the autumn of 2008. Most of it was pre-fabricated, which enabled the builders to shorten the construction time and simplified the construction work on-site as well. This technology combines the pros of a high degree of prefabrication and a low transport weight with a short time of installation. VFT-WIB has taken all the advantages of knowledge gained in the VFT technology research into composite beams and combined them successfully with the traditional technology of encased beams. The new technology brings along the productive application of composite steel and concrete structures that are manufactured as prefabricated segments. A VFT-WIB beam makes use of a variety of methods of composite action using specially-adjusted strip connectors. Ideally, the T-shaped steel beams are made of a single rolled section. An I-section is flame-cut longitudinally, whereas the perfect and exact geometry of the section must be maintained, which divides the section into two symmetrical units (Fig. 1.9). The precision of cutting the section while using as much material as possible together with the suitable combination of composite action between steel and concrete presents the future orientation in the development of VFT beams [14, 15].

The specially-adjusted steel T-sections, as shown in Fig. 1.10, can make new types of sections with high shear resistance of composite members. The bare steel

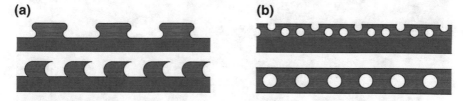

Fig. 1.9 a Symmetrical cutting of rolled sections b various shapes of strip connectors

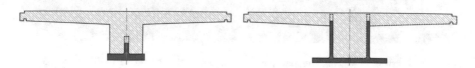

Fig. 1.10 Cross-section of a VFT-WIB beam with a single or double T-section

component of the beam exposed to weather must be surfaced before placed in concrete.

Steel T-sections are laid in a form, reinforced and encased in concrete in the form prepared. After hardening, the beams are transported to a construction site and placed on bearings. Subsequently, the concrete deck is made monolithic without using any additional formwork. The cross-sectional shape of the beam combines all significant advantages regarding the durability and economy of the structure. VFT-WIB beams can find their application in both road and railway bridges. Excellent results have so far been shown in the construction of new railway bridges, replacing frequently used deck bridges with encased filler-beams. The majority of the old bridges are about to end their operable life; therefore, they need reconstruction or full replacement. New bridges that are to replace them will have to fulfil the following requirements:

- The slenderness of a load-bearing structure and low production costs.
- Adjustability during the replacement of the existing deck bridges.
- Minimum environmental impact.
- Resistance and durability.

VFT-WIB sections can meet the above requirements to a great extent. Prefabricated beams can thus be effectively employed in the replacement of bridges on main railway routes, compliant with the necessity of shrinking the time of traffic closure to the minimum. Additionally, VFT-WIB technology allows the building of simple temporary bridges cost-effectively, and the steel beam is subject to only minimal fatigue [12, 16] (Fig. 1.11).

Fig. 1.11 A bridge in Vigaun, Austria, spanning 26.15 m sections with VFT-WIB beams. *Source* ÖBB Österreichische Bundesbahnen, ArcelorMittal Company

Having further developed the VFT-WIB beams for railway transport, *Max Bogl* designed a VFT-RAIL composite steel and concrete beam VFT-RAIL. The beam consists of a pair of steel T-sections inverted against each other with additional reinforcement bars and stirrups. Rails are fastened directly down onto the beam without any ballast bed. This way, a fixed position of the rails for high-speed tracks can be ensured. The exceptional stiffness of the cross-section can guarantee their strong fatigue resistance [2, 13, 17].

1.6 Composite Pre-flexed Beams

Another type of patented composite construction beams is pre-flexed beams, i.e. pre-stressed beams that were initially designed in Belgium in 1951. Rolled wide-flange beams, reinforced with additionally-welded flanges if necessary, are elastically bent with a pressing jack, and in this pre-stressed condition, a concrete deck is cast only around the bottom tension flanges. When the concrete has set, the load is released, the beams straighten up, and the concrete surrounding the bottom flanges is compressed. The top flanges and the webs in the deck are filled with concrete on-site. This type of pre-flexed beam and the process of its construction were developed by A. Lipski in close collaboration with L. Baes [18].

The manufacturing process is as follows (Fig. 1.12).

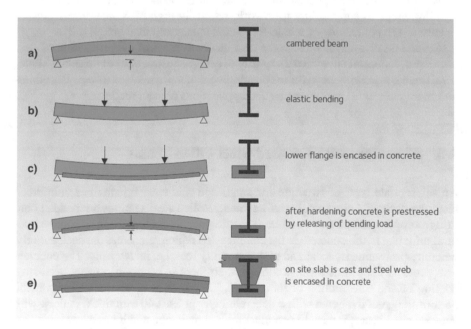

Fig. 1.12 Process of manufacturing a pre-flexed beam. *Source* ArcelorMittal Company

This type of beam is suitable when there is little clear headway under a bridge, or there is little handling room for the renewal of coating on the steel components of a bridge, for example, in bridges over electrified rail tracks.

For economic reasons, it is desirable to produce pre-flexed beams in series. The economy is also increased since the pair of beams pre-stressed against each other is manufactured simultaneously. However, when compared to other types of composite beams, the production of this type of beam is still more complicated and costlier.

By pre-stressing a beam, its load-bearing capacity will not increase, yet its deflection due to variable load will decrease. As a result, such beams remain very slender. In railway bridges, the depth-to-span ratio of a beam varies from 1/25, and in road bridges, it is as high as 1/40. Stress fluctuations caused by a variable load are smaller than in other types of beams, and so is the deterioration of its mechanical properties due to fatigue. Pre-compression of the concrete around the bottom flange reduces the resulting tensile stress in a beam and thus also reduces cracking in the concrete [2, 19].

1.7   Characteristics of Deck Bridges

Decks with encased filler-beams are suitable for short and mid-span bridges. They have many advantages such as small headway, a familiar behavioural pattern, straightforward structural design, fast and simple construction without the necessity of building props to support any formwork, relatively high resistance to car crashes in case of vehicular bridges, hassle-free maintenance and many more [20].

Several disadvantages include the cost ineffectiveness of the production of steel I-sections, commonly used in the majority of bridge constructions of this kind. Therefore, there is a growing need for the development of more relevant design procedures and more purposeful arrangements and applications of steel sections.

1.8   Structures with Encased Steel Filler-Beams

An appropriate type of structure for small and mid-span bridges is a reinforced concrete deck with rigid steel reinforcement. Rolled steel sections are bedded onto bridge supports and cast in concrete. The components act together as a single structural unit; that is, the concrete in the compression region reinforces the deck entirely, whereby preventing its local and global stability loss, i.e. its *buckling*. The concrete also protects the beams against corrosion, so there is no need to use a protective coating on the construction or its renewal. Therefore, it is advisable to cover the bottom flange of the beam with a concrete layer at least 40 mm thick. A deck with encased steel beams is easy to manufacture and maintain, and it has a low structural height. One of the drawbacks is relatively high steel consumption resulting from the

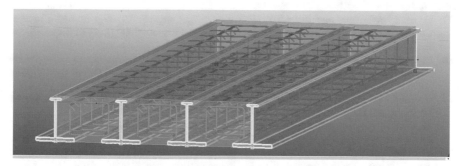

Fig. 1.13   A deck structure with encased steel I-sections

symmetry of the rolled section and the necessity to place and arrange the sections relatively densely inside the concrete deck [21, 22] (Fig. 1.13).

References

1. J. Bujňák, *Kovové konštrukcie a mosty* (Vysoká škola dopravy a spojov v Žiline, 1995), 217 pp. ISBN 80-7100-262-3
2. R. Zanon, N. Popa, *ArcelorMittal Belval & Differdange, Bridges with Rolled Sections, Advanced Solutions for Rolled Beams in Bridge Construction* (2018)
3. Státní ústav dopravního projektování v Praze, *Směrnice pro návrh a provádění ocelobetonových nosních konstrukcí železničních mostů* (1981)
4. Státní ústav dopravního projektování Praze, *Železniční deskové mostní konstrukce se zabetonovanými ocelovými nosníky* (1971)
5. STN EN 1994-2, *Navrhovanie spriahnutých oceľobetónových konštrukcií, časť 2:Všeobecné pravidlá a pravidlá pre mosty*
6. STN EN 1994-1-1, *Eurokód 4: Navrhovanie spriahnutých oceľobetónových konštrukcií. Časť 1–1: Všeobecné pravidlá a pravidlá pre budovy*
7. L. Hubinská, *Optimalizácia priečneho rezu konštrukcií doskových mostov so zabetónovanými nosníkmi* (Žilina, 2007)
8. V. Kvočák, V. Kožlejová, Research into filler-beam deck bridges with encased beams of various sections. Tech. Gaz. **18**(3), 385–392. ISSN 1330-3651
9. Technická správa: Trať Michaľany-Palota, oprava mosta v km 5 (2009)
10. V. Kožlejová, *New and Traditionally Used Manners for Shear Connection in the Composite Steel-Concrete Structures.* Mladý vedec, 2009. Podbanské, 15–16 April 2009
11. G. Seidl, *Prefabricated Enduring Composite Beams Based on Innovative Shear Transmission,* 1st July 2006 to 30th June 2009 (E. ViefhuesArcelor Profil Luxembourg S.A., Luxembourg (ARCELOR)).
12. A. Ďuricová, M. Rovňák, *Navrhovanie oceľovo-betónových konštrukcií podľa STN EN 1994-1-1* (VEDA vydavateľstvo SAV, Bratislava, 2008). ISBN 978-80-224-1022-9
13. https://www.cticm.com/sites/default/files/06_gs_exemples_ponts_precobeam.pdf
14. https://www.arcelormittal.com/sections/fileadmin/redaction/4-Library/1
15. G. Seidl, *Starting from VFT Getting External Reinforcement* (SSF Ingenieure AG, Berlin, 2013)
16. O. Hechler, L.-G. Cajot, P.-O. Martin, A. Burceau, Efficient and economical design of composite bridges with small and medium spans, in *7th International Conference of Steel Bridges,* Guimaraes, Portugal, 4–6 June 2008

17. G. Seidl, *Behaviour and Load-Bearing Capacity of Composite Dowels in Steel-Concrete Composite Girders*. Rozprawa doktorska. Raport serii PRE nr 4/2009 (Instytut Budownictwa Politechniki Wrocławskiej, 2009)
18. PreCo-Beam, *Prefabricated Enduring Composite Beams Based on Innovative Shear Transmission*. Research Fund for Coal and Steel, Contract N° RFSR-CT-2006-00030. 01/07/2006–30/06/2009
19. G. Seidl, O. Hoyer, *VFT Bauweise Verbund-Fertigteil-Bauweise* (SSF Ingenieure AG, Munchen, 2011).
20. M. Rovňák, A. Ďuricová, K. Kundrát, Ľ. Naď, *Spriahnuté oceľovo-betónové mosty* (Košice, Elfa s. r. o., 2006). ISBN 80-8073-485-2
21. M. Fragiacomo, C. Amadio, L. Macorini, Influence of viscous phenomena on steel-concrete composite beams with normal or high-performance slabs. Steel Compos. Struct. **2**(2), 85–98 (2002)
22. Š. Gramblička, J. Bujňák, V. Kvočák, J. Lapos, *Zavádzanie Eurokódov do praxe: Navrhovanie spriahnutých oceľobetónových konštrukcií STN EN 1994-1-1* (Inžinierske konzultačné stredisko Slovenskej komory stavebných inžinierov, Bratislava, 2007). ISBN 978-80-89113-36-1

Chapter 2
Proposed Designs of Deck Bridges

The currently designed and constructed deck bridges with encased filler-beams predominately contain rolled or welded steel I-sections. The efforts to better employ the steel section have led the authors to the idea of designing a new steel section that could act mainly in the tension region of a future composite bridge structure. Different types of steel section have been considered with the goal to design and experimentally verify deck bridges with modified steel sections and achieve major economies, chiefly by reducing the amount of steel for their production [1].

In the laboratories of the Institute of Structural Engineering at the Faculty of Civil Engineering at the Technical University of Košice, the first series of beam specimens with modified shapes of steel section was developed and tested, and the resistances of a full steel I-section and an inverted T-section were compared. The individual variants differed in the method of ensuring composite action between the steel and concrete. Smooth and comb-like web edges of the T-section were also compared (Fig. 2.1).

2.1 Preliminary Results of the First Measurements

In order to determine the resistance and composite action of newly-shaped encased steel sections, several experimental measurements have been taken on beams with T-sections. The measurements have proven the assumption that the reduction in steel consumption would not lead to a reduction in the resistance and stiffness of the composite member. The results have also indicated that special attention should be placed on the method of ensuring shear connection in the member, which may be crucial when deciding on the application of alternative types of beams.

Variables were continuously measured and recorded, and the average values were graphically depicted and evaluated. The correlation between the overall mid-span deflection and the amount of load applied is shown in Fig. 2.2. Similarly, the

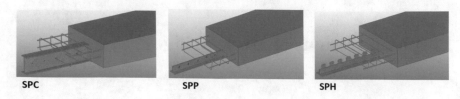

Fig. 2.1 Modified steel sections in a composite beam

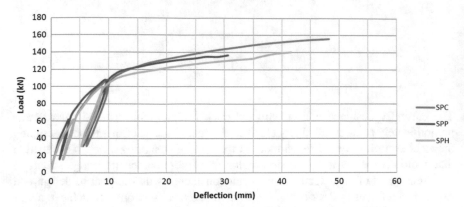

Fig. 2.2 Correlation between the deflection and the amount of load applied to a beam. *Source* Kožlejová [2]

strains/relative deformations in the concrete and steel cross-sections were evaluated and displayed in Figs. 2.3 and 2.4 [2, 3].

During the observation of strains in the steel and concrete members, the method providing the composite action between the steel T-sections with straight, smooth

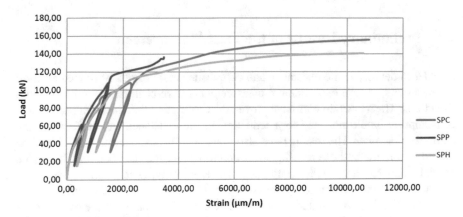

Fig. 2.3 Correlation between the strain in a steel section and the load applied to a beam. *Source* Kožlejová [2]

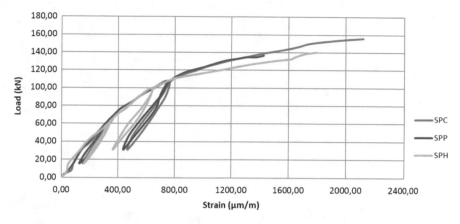

Fig. 2.4 Correlation between the strain in a concrete deck and the load applied to a beam. *Source* Kožlejová [2]

edges and the concrete has appeared to be problematic. As can be seen from the diagrams above, although the resistance of the modified beams is satisfactory, it is still necessary to improve how the steel and concrete members act compositely together. This can be done either by supplementing the beam with shear connectors or densifying the transverse reinforcement.

Deck bridges are frequently cast on site. The bottom flange becomes part of the so-called *false formwork* or *falsework*. Before the concrete deck is set and hardened, steel beams carry the full weight of the structure. Steel T-sections, however, do not have sufficient strength during the construction process, which means that the main advantage of this type of deck bridge, being it fast in situ construction without using props, is lost. Hence, T-sections can be employed in pre-cast members. When designing such members, it is always necessary to account for their final position in a structure and their final installation on site.

References

1. R. Vargová, K. Chupayeva, P. Beke, *Príprava experimentálneho programu mostov so zabetóno-vanými oceľovými nosníkmi modifikovaných tvarov* (Vedecko-výskumná činnosť ÚIS, Herľany, 2012)
2. V. Kožlejová, New and traditionally used manners for shear connection in the composite steel-concrete structures. *Mladý vedec* (Podbanské, 2009)
3. V. Kvočák, V. Kožlejová, Research into filler-beam deck bridges with encased beams of various sections. Tech. Gaz. **18**(3), 385–392 (2011). ISSN 1330-3651

Chapter 3
The Tests of Deck Bridges with Encased Steel Beams

3.1 Design of Laboratory Specimens

Drawing on the knowledge gained in the previous research studies, a further number of modified shapes of steel sections for composite members have been designed in the Institute of Structural Engineering at the Faculty of Civil Engineering at the Technical University of Kosice.

The experimental research programme included tests on five types of real-size experimental specimens with encased steel connecting strips, providing different methods of composite action. The dimensions of the specimens were designed with respect to the conditions available at the Laboratory of Excellent Research of the Civil Engineering Faculty. Therefore, the length and the width of all specimens were 6,000 mm and 900 mm, respectively. Their depth arose from the previous dimensions and the proportional arrangement and positioning of the steel strips within the actual composite members, which was 270 mm. There were two steel sections in each specimen. Longitudinally, the sections were supplemented with reinforcing bars 12 mm in diameter on the edges, securing their stability and interaction with transverse reinforcement.

Five different types of composite members were manufactured for experimental verification. N1 beams were made with steel box sections filled with concrete. The steel box section was created by welding a 6-mm-thick U-shaped steel sheet, forming the top flange and webs of the section, to another 6-mm-thick steel sheet, forming the bottom flange with overhanging ends. Holes 50 mm in diameter were flame-cut in the webs at an axial distance of 100 mm. Transverse reinforcement bars 12 mm in diameter were threaded through every third hole in the beam. Similarly, the top flange was perforated by holes 50 mm in diameter at an axial distance of 100 mm. The holes were arranged in an alternative manner so that they appeared either in the webs or in the flange of each section. The cross-section and longitudinal section of beam N1 are given in Fig. 3.1.

© The Author(s), under exclusive license to Springer Nature Switzerland AG 2021
V. Kvočák and D. Dubecký, *Research and Development of Deck Bridges*,
SpringerBriefs in Applied Sciences and Technology,
https://doi.org/10.1007/978-3-030-66925-6_3

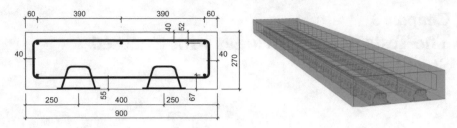

Fig. 3.1 Composite beam N1

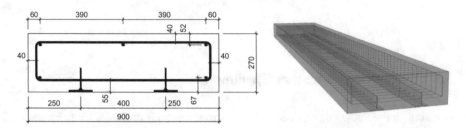

Fig. 3.2 Composite beam N2

N2 beams, as shown in Fig. 3.2, were made with encased steel T-sections. The T-sections were created by cutting rolled IPE 220 sections straight longitudinally. Transverse reinforcement bars 12 mm in diameter were threaded through 20 mm holes made in the webs at an axial distance of 100 mm. The holes were flame-cut 55 mm from the bottom edge.

Another type of specimen is N3 beams displayed in Fig. 3.3. Similarly to the previous specimens, these were also made with encased steel T-sections flame-cut from rolled IPE 220 sections, but with comb-like edges. The shape and dimensions of these beams follow the design of the previously tested connecting strip at the Faculty of Civil Engineering at the Technical University of Kosice, which also made it possible to calculate the load-bearing capacity of the composite member theoretically.

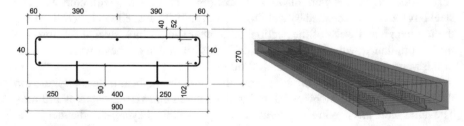

Fig. 3.3 Composite beam N3

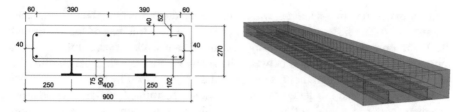

Fig. 3.4 Composite beam N4

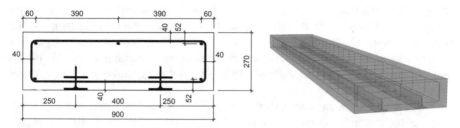

Fig. 3.5 Composite beam N5

Transverse reinforcement was laid into each tooth of the comb, which was axially every 105 mm at the height of 55 mm.

N4 beams, as shown below in Fig. 3.4, were also made from rolled steel IPE 220 sections, where the web edges were purposefully adjusted perforated strips, providing improved composite action between the steel and concrete members. The web edge was modified with holes flame-cut in two rows at an axial distance of 90 mm. Transverse reinforcement was also positioned at two levels at a distance of 135 mm [1].

Finally, the last type of composite beam is N5 beams displayed in Fig. 3.5. Like N2 beams, these were made from encased T-sections created by flame-cutting rolled steel IPE 220 sections in half. Transverse reinforcement bars 12 mm in diameter were placed as high as 40 mm from the bottom at an axial distance of 300 mm. Composite action was ensured using loops made from rebars with the dimensions of 50 × 100 mm. The loops were welded and attached to the steel section webs horizontally at the height of 60 mm measured from the bottom edge.

3.2 Theoretical Analysis of the Composite Member Resistance

This chapter deals with the rigid steel reinforcement of a steel box section in more detail.

This type of specimen is marked as N1 beams in the previous sections.

In comparison with T-sections, which seem to be the most cost-effective, the steel box section forms falsework during the process of construction of the beam and allows the omission of the supporting system of formwork props. Furthermore, it improves the composite action between the steel and concrete members within the beam, so that slip between them is eliminated during the loading stage.

Both the webs and top flanges of the beams are perforated, and transverse reinforcement bars are threaded through the openings in the webs. The primary rigid reinforcement in the experimental specimen N1 is designed as a welded steel box section. The cross-section and longitudinal section of the N1 specimen are provided in Fig. 3.6. Unlike traditional I-sections, which find the same applications, beams with encased steel box sections manifest higher bending resistances during their service life, i.e. after the shear connection has become effective, and they act compositely in the finalised structural members. A parametric chart comparing the N1-N5 beams and various methods of composite action is provided at the end of each experiment (Fig. 3.7).

Based upon the theoretical analysis, the preliminary assumed pure bending resistance of the specimens was calculated. An assumption was made for the plastic behaviour of the steel sections. Two cross-sections were evaluated to decide the bending resistance: cross-section 1, running through the centres of gravity of the holes in the webs, and cross-section 2, running through the centre of gravity of the hole in the flange of the steel box section.

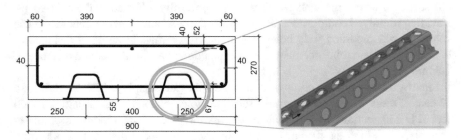

Fig. 3.6 A steel box section in the composite beam N1

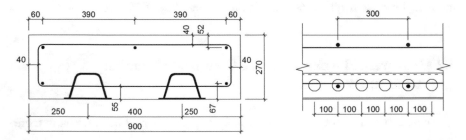

Fig. 3.7 Composite beam N1 and the positioning of holes in the steel box section webs

Material characteristics:

$f_y = 315.3$ MPa, the nominal value of the yield strength of structural steel.
$f_{ck} = 32.17$ MPa, the characteristic value of the cylinder compressive strength of concrete.
$f_{sk} = 490$ MPa, the characteristic value of the yield strength of reinforcing steel.
$f_{ctk} = 1.5$ MPa.
$E_c = 32$ GPa, the modulus of elasticity for concrete.
$\gamma_a = 1.0$, a partial factor for steel.
$\gamma_c = 1.5$, a partial factor for concrete.
$\gamma_{as} = 1.15$, a partial factor for reinforcing steel.

Cross-sectional characteristics:

Cross-section 1 (Fig. 3.8):

$b = 900$ mm, the width of the specimen; $b/2$ is considered in the calculations.
$h = 270$ mm, the depth of the specimen.
$A_{a,1} = 2{,}438.94$ mm², the cross-sectional area of the structural steel section.
$A_s = 113$ mm², the cross-sectional area of a reinforcement bar.
$a_s = 67$ mm, the position/distance of the centre of gravity of the reinforcement bar from the bottom edge.
$z_{a,1} = 42.29$ mm, the position/distance of the centre of gravity of the steel section from the bottom edge.
$b_a = 200$ mm, the width of the flange of the steel section.
$h_a = 110$ mm, the depth of the structural steel section.

The position of the plastic neutral axis is given by:

$$N^- = N^+ \tag{3.1}$$

$$b \cdot z \cdot 0.85 \cdot \frac{f_{ck}}{\gamma_c} = A_a \cdot \frac{f_y}{\gamma_a} + A_s \cdot \frac{f_{sk}}{\gamma_{as}}, \tag{3.2}$$

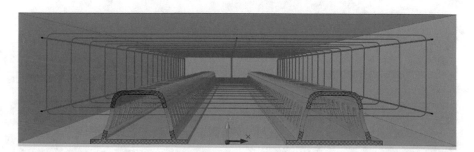

Fig. 3.8 Section 1—holes in the webs

resulting in $z = 0.066994$ m.

Consequently, the design value of the plastic resistance moment of the composite section with full shear connection is determined from the expression as follows:

$$M_{pl,Rd} = b \cdot \frac{z^2}{2} \cdot 0.85 \cdot \frac{f_{ck}}{\gamma_c} + A_a \cdot (h - z_a - z) \cdot \frac{f_y}{\gamma_a} + A_s \cdot (h - z - a_s) \cdot \frac{f_y}{\gamma_a}$$

$$\text{(3.3)}$$

$$0.85 \cdot \frac{f_{ck}}{\gamma_c} = 18.229 \, \text{MPa} \tag{3.4}$$

$$\frac{f_y}{\gamma_a} = 315.3 \, \text{MPa} \tag{3.5}$$

The theoretical value of the plastic resistance moment of the composite section is then calculated as $M_{pl,teor,1} = 317.48$ kNm.

Cross-section 2 (Fig. 3.9):

$b = 900$ mm, the width of the specimen.
$h = 270$ mm, the depth of the specimen.
$A_{a,2} = 2{,}738.94 \, \text{mm}^2$, the cross-sectional area of the structural steel section.
$A_s = 113 \, \text{mm}^2$, the cross-sectional area of the reinforcement bar.
$a_s = 67$ mm, the position/distance of the centre of gravity of the reinforcement bar from the bottom edge.
$z_{a,2} = 37.93$ mm, the position/distance of the centre of gravity of the steel section from the bottom edge.
$b_a = 200$ mm, the width of the flange of the steel section.
$h_a = 110$ mm, the depth of the structural steel section.

The position of the plastic neutral axis is given by the following expression (3.6):

$$N^- = N^+ \tag{3.6}$$

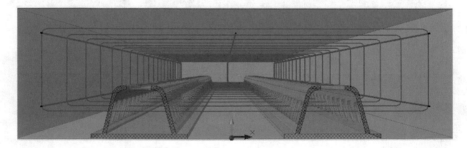

Fig. 3.9 Section 2—holes in the top flange

$$b \cdot z \cdot 0.85 \cdot \frac{f_{ck}}{\gamma_c} = A_a \cdot \frac{f_y}{\gamma_a} + A_s \cdot \frac{f_{sk}}{\gamma_a} \tag{3.7}$$

$$z = 0.0746815 \, \text{m}$$

Consequently, from the expression (3.3), the theoretical value of the plastic resistance moment of the composite section is calculated as

$$M_{pl,teor,2} = 357.62 \, \text{kNm}$$

As a result, the cross-section that will decide the load-bearing capacity of the entire composite member, according to the theoretical value of the plastic resistance moment, is cross-section 1 (Fig. 3.8), with $M_{pl,teor,1} = 317.48$ kNm.

Based on the calculated characteristic value of the plastic resistance moment and the loading scheme, it is possible to determine the assumed forces at which the specimen should fail. The values of the forces necessary for the specimen to reach the ultimate resistance can be specified from the expression (3.8) as follows:

$$M_{pl,Rk} = F_{pl,Rk} \cdot x + \frac{1}{8} \cdot g \cdot l^2 \tag{3.8}$$

where $l = 5.8$ m is the theoretical span of the specimens and
$x = 2.0$ m is the position/distance of the force from the nearest support.

The given value of the resistance moment corresponds to the loading force $F_{pl,teor,2} = 145.08$ kN.

Next, the mid-span deflection is determined as

$$\delta = \frac{F_{pl,Rd} \cdot x}{24 \cdot EI_i} \left(3 \cdot l^2 - 4 \cdot x^2 \right) \tag{3.9}$$

where l is the span of the specimens and
x is the position/distance of the force from the nearest support.
Upon the numerical substitution:

$$\delta = \frac{145.08 \cdot 2}{24 \cdot 12715.97} \left(3 \cdot 5.8^2 - 4 \cdot 2^2 \right)$$

$$\delta = 80.75 \, \text{mm}$$

The average values of the deflections measured in the specimens, corresponding to the ultimate load that has been considered in the calculations, as well as the percentage comparison with the deflections previously calculated, are provided in Table 3.1.

Table 3.1 Comparison of the deflections measured and calculated for the individual specimens

Specimen	δ_{exp} (mm)	Difference (%)	$\delta_{exp, average}$ (mm)	$F_{pl,Rd}$ (kN)	δ (mm)	Difference (%)
N1-1	75.20	7.38	76.83	145.08	80.75	5.1
N1-2	76.70	5.28				
N1-3	78.60	2.73				

For deck bridges with encased steel beams, the STN EN 1994-2 standard recommends considering flexural stiffness as the average of the value of the stiffness of the uncracked section and the value of the stiffness of the cracked section with the maximum crack length. Calculations have shown, however, that in this particular case, it is more accurate to use the cracked flexural stiffness for the whole specimen. This is because the specimens were loaded in a manner causing the maximum bending moment approximately at 1/3 of the specimen length, close to the middle—and at the same time, the design value of the plastic resistance moment was achieved all along this specimen segment. Flexural stiffnesses EI_i were considered in the calculation of deflections in the specimens [2, 3].

Cracks arise in the concrete region where tensile stresses exceed the tensile strength of concrete; therefore, the cracked section must be eliminated from the calculations of the flexural stiffness of the cross-section. The tensile strength of concrete is low, and the uncracked concrete in tension is very close to the centre of gravity of the cross-section; therefore, its contribution to the flexural stiffness of the cross-section can be neglected and eliminated from the calculations. The position of the centre of gravity of the ideal cross-section is not definitely or unambiguously specified. It can be determined iteratively, using the following recurrent formula [4, 5]:

$$z_{i,(N+1)} = \frac{A_a \cdot (h - z_a) + \frac{b \cdot z_{i,(N)}^2}{2 \cdot n_0}}{A_a + \frac{b \cdot z_{i,(N)}}{n_0}} \quad (3.10)$$

The first step assumes that the neutral axis is not positioned in the structural steel section, therefore,

$$z_{i(N=1)} = h_c$$

The iterative process finishes with a definite condition:

$$\left| z_{i,(N)} - z_{i,(N+1)} \right| \leq 0.001, \quad \text{then } z_i = z_{i,(N+1)} \quad (3.11)$$

The ratio of the moduli of elasticity is given by a modular ratio

$$n_0 = \frac{E_a}{E_c} = 6.5625 \quad (3.12)$$

The following moduli of elasticity were taken into account for the specification of the flexural stiffness of the composite beam:

$E_a = 210,000$ MPa for the steel members and
$E_c = 32,000$ MPa for the concrete members.

Having substituted the numerical values, it follows that:

$$z_{i,(N+1)} = \frac{2438.94 \cdot (270 - 42.29) + \frac{450 \cdot z_{i,(N)}^2}{2 \cdot 6.5625}}{2438.94 + \frac{450 \cdot z_{i,(N)}}{6.5625}} \quad z_{i,1} = 93.39 \text{ mm.} \tag{3.13}$$

Furthermore, the flexural stiffness EI_i of the composite beam is provided by the following expression:

$$EI_i = E_a \cdot \left(I_a + A_a \cdot (h - z_i - z_a)^2\right) + E_c \frac{b \cdot z_i^3}{3} \quad EI_i = 12715.97 \text{ kPa m}^4 \tag{3.14}$$

The deflection caused by the effects of creep of concrete

Deformation due to creep of concrete is calculated from the creep coefficient and the average secant modulus of elasticity, and it is given as:

$$\varepsilon_{cc}(\infty, t_0) = \varphi(\infty, t_0) \cdot (\sigma_c / 1.05 \, Ecm),$$

where

$\varepsilon_{cc}(\infty, t_0)$ is the creep deformation at time $t = \infty$
$\varphi(\infty, t_0)$ is the creep coefficient at time $t = \infty$
σ_c is the constant compressive stress applied at time $t = \infty$.

Appropriate allowances shall be made for the effects of non-linear creep in cases when the compressive stress in the concrete at the age of t_0 is higher than $0.45 \, f_{ck}(t_0)$. Then, the creep deformation is determined by the following expressions [4, 6]:

$$\varepsilon_{cc} = \varphi(t, t_0) \cdot \frac{\sigma_c}{E_c} \quad \varepsilon_{cc} = 1.40 \cdot \frac{13.50}{33000} = 5.72 \cdot 10^{-4} \tag{3.15}$$

$$\sigma_c = 0.45 \cdot f_{ck} \quad \sigma_c = 0.45 \cdot 30 = 13.5 \text{ MPa} \tag{3.16}$$

The creep coefficient $\varphi(t, t_0)$ is calculated from the formula as follows:

$$\varphi(t, t_0) = \varphi_0 \cdot \beta_c(t, t_0) \quad \varphi(t, t_0) = 1.931 \cdot 0.725 = 1.40 \tag{3.17}$$

φ_0 is a theoretical creep coefficient and, compliant with the standard in question, it is allowed to be estimated by:

$$\varphi_0 = \varphi_{RH} \cdot \beta_{fcm} \cdot \beta(t_0) \quad \varphi_0 = 1.450 \cdot 2.725 \cdot 0.488 = 1.931 \tag{3.18}$$

The coefficient that considers the effects of relative humidity on the creep coefficient is determined by the expression:

$$\varphi_{RH} = \left[1 + \frac{1 - RH/100}{0.1 \cdot \sqrt[3]{h_0}} \cdot \alpha_1 \right] \cdot \alpha_2 = \left[1 + \frac{1 - 80/100}{0.1 \cdot \sqrt[3]{337.5}} \cdot 0.944 \right] \cdot 0.983 = 1.450$$
$$\tag{3.19}$$

while the ambient relative humidity is taken as $RH = 80\%$.

A substitutional depth of the member is calculated according to the formula:

$$h_0 = \frac{2 \cdot A_c}{u} = 337.5 \, \text{mm} \tag{3.20}$$

The factors by which the effect of the mean value of the measured cylinder compressive strength of the concrete is included are:

$$\alpha_1 = \left[\frac{35}{fcm} \right]^{0.7} = \left[\frac{35}{38} \right]^{0.7} = 0.944 \quad \alpha_2 = \left[\frac{35}{fcm} \right]^{0.2} = \left[\frac{35}{38} \right]^{0.2} = 0.983$$

The factor β_{fcm}, which accounts for the effect of the value of the compressive strength of the concrete on the theoretical creep coefficient, is given by the following formula:

$$\beta_{fcm} = \frac{16.8}{\sqrt{fcm}} \quad \beta_{fcm} = \frac{16.8}{\sqrt{38}} = 2.725 \tag{3.21}$$

$$\beta(t_0) = \frac{1}{\left(0.1 + t_0^{0.2} \right)}$$

$$t_0 = 28 \, days \quad \beta(t_0) = \frac{1}{\left(0.1 + 28^{0.2} \right)} = 0.488 \tag{3.22}$$

The factor that describes the development of creep effect in time, from the initial moment of loading the composite member, can be estimated using the following expression:

$$\beta_c(t, t_0) = \left[\frac{(t - t_0)}{(\beta_H + t - t_0)} \right]^{0.3} \tag{3.23}$$

$$\beta_c(t, t_0) = \left[\frac{(420 - 28)}{(751.96 + 420 - 28)} \right]^{0.3} = 0.725$$

The factor that depends on the ambient relative humidity RH (%) and the substitutional depth of the composite member is determined by the expression:

$$\beta_H = 1.5 \cdot \left[1 + (0.012 \cdot RH)^{18}\right] \cdot h_0 + 250 \cdot \alpha_3 \leq 1500 \cdot \alpha_3 \tag{3.24}$$

$$\beta_H = 1.5 \cdot \left[1 + (0.012 \cdot 80)^{18}\right] \cdot 337.5 + 250 \cdot 0.959 \leq 1500 \cdot 0.959$$
$$751.96 \leq 1439.57, \text{ satisfying the standard condition.}$$

The factor by which the effect of the mean value of the measured cylinder compressive strength of the concrete is included is:

$$\alpha_3 = \left[\frac{35}{38}\right]^{0.5} = 0.959 \tag{3.25}$$

The resulting stress due to the effects of creep of concrete is then:

$$\sigma_{c,m} = \varepsilon_{cc} \cdot E_c = 5.72 \cdot 10^{-4} \cdot 33000 = 18.91 \, \text{MPa} \tag{3.26}$$

The value of the compressive normal force on the composite member due to the effects of creep of concrete is determined by the expression:

$$N_{cc} = \sigma_{c,m} \cdot A_c = 18.91 \cdot 4.86 \cdot 10^3 = 91.9 \, \text{kN} \tag{3.27}$$

The resulting bending moment caused by the effects of creep of concrete is determined as:

$$M_{cc} = N_{cc} \cdot r_c = 91.9 \cdot 0.270 = 24.8 \, \text{kNm} \tag{3.28}$$

The resulting deflection caused by the effects of creep of concrete

$$w_{\text{max},c} = \frac{5}{48} \cdot \frac{(M_{cc}) \cdot L^2}{\left(\frac{E_0}{n_0}\right) \cdot I_0 \cdot n} = 4.19 \, \text{mm} \tag{3.29}$$

The maximum deflection caused by the effects of creep of concrete occurs at mid-span of the composite member and accounts for 4.19 mm.

3.3 Preparation of Laboratory Tests of Composite Beams

Following the theoretical assumptions and calculations, several N1-type experimental specimens with encased steel box sections were prepared in the laboratories of the Institute of Structural Engineering at the Faculty of Civil Engineering at the

Technical University of Kosice. The specimens were cast into steel forms placed onto bedding so that sagging (deflection) of the composite beams was prevented during concrete placement. Steel box sections and reinforcement bars were supplied by specialised companies in the shapes and dimensions required. To make sure that the concrete for all designed specimens was the same quality, concrete Class STN EN 206–1—C30/37–XC4, XD3, XF4, Cl 0, 4—Dmax16 was delivered from a nearby central mixing plant.

Three beam specimens were made at the time as well as specimens to verify the actual material properties of the concrete. After the concrete had set and hardened, the forms were removed, and the whole cycle continued until all experimental specimens for static, long-term, and fatigue tests were made. The manufacturing process of composite members with rigid reinforcement took place in four steps as follows (Fig. 3.10):

3.3.1   Preparation of Steel Sections

All steel sections were manufactured and supplied according to the design specifications, in the shapes and dimensions required. Measurement gauges were fastened to the rigid steel reinforcement in the stage following the manufacture. These were located in the middle of the beam span adjacent to the holes perforated in the webs and the top flange, as displayed in Fig. 3.11. Strain gauges were also attached to the bottom flange after the removal of the forms. All strain gauges were bonded to the clean and degreased surface of the steel sections using two-component epoxy glue. After the glue had hardened, electric wires were soldered onto the sections for signal monitoring.

3.3.2   Binding Reinforcement Bars

Bent reinforcement bars were supplied by a specialised company according to the submitted design specifications. Stirrups and longitudinal reinforcement bars were bound by spot welds at several points. Stirrups were threaded through the holes in the webs of the rigid steel reinforcement during the welding process, as shown in Fig. 3.12. The spacing between the stirrups was 300 mm. Longitudinal reinforcement bars were positioned at all four corners of the stirrups, and an extra bar was added to divide the reinforcement in the middle at the upper edge.

1. Preparation of steel sections

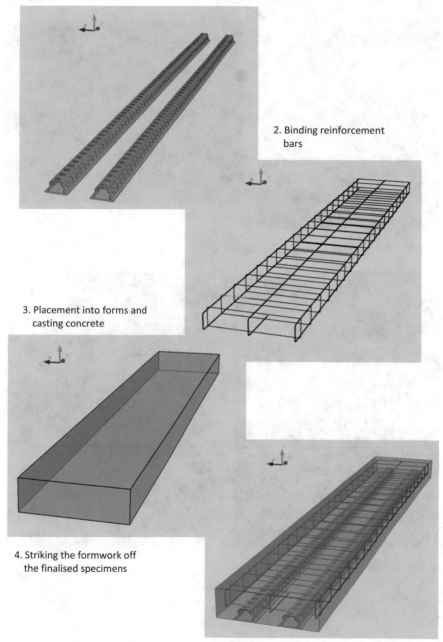

2. Binding reinforcement
   bars

3. Placement into forms and
   casting concrete

4. Striking the formwork off
   the finalised specimens

Fig. 3.10 Process of manufacturing composite members

Fig. 3.11 Gluing and connecting strain gauges

Fig. 3.12 Binding reinforcement bars

Fig. 3.13 Manufacturing the experimental specimens

3.3.3 Placement into Forms and Casting Concrete

The ready-made reinforcement cages together with the rigid steel reinforcement, i.e. steel box sections, were placed into prepared steel forms, where release agent had been previously applied to ensure that they separate quickly and cleanly from the concrete surface. After the steel reinforcement had been centred and fixed, concrete was cast into the forms and vibrated. The surface of the beams was levelled and finally manually smoothed with a trowel. Fresh samples of concrete were taken during the concreting process for the verification of their actual material properties (Fig. 3.13).

3.3.4 Striking the Formwork off the Finalised Specimens

All specimens were stored for a minimum of 28 days after the concrete had hardened and the formwork had been stripped off. In fact, the process of striking off the forms had meant releasing the screws all around the perimeter of the steel forms. An overhead mobile crane and a pallet mover were used for handling the specimens. The entire manufacturing cycle repeated until the sufficient number of experimental specimens was produced (Fig. 3.14).

Fig. 3.14 Specimens in the phase of hardening and handling with finish sample

References

1. A. Li, K. Cederwall, Push-out tests on studs in high-strength and normal-strength concrete. J. Constr. Steel Res. **36**(1), 15–29 (1996)
2. STN EN1991-2 Eurokód 1, Zaťaženia konštrukcií, casť 2: Zaťaženia mostov dopravou
3. STN EN 1991-3 Eurokód 1, Zásady navrhovania a zaťaženia konštrukcií, časť 3: Zaťaženia mostov dopravou
4. STN EN 1992-1-1 Eurokód 2, Navrhovanie betónových konštrukcií, časť 1-1: Všeobecné pravidlá a pravidlá pre pozemné stavby
5. STN EN 1994-2, Navrhovanie spriahnutých oceľobetónových konštrukcií, časť 2: Všeobecné pravidlá a pravidlá pre mosty
6. F.S.K. Bijlaard, G. Sedlacek, J.W.B. Stark, Procedure for the determination of design resistance from tests. Delft (1988)

Chapter 4
Measurement of Material Properties of Concrete and Steel

In order to specify the material properties of the concrete used in experimental members, several laboratory tests have been carried out, namely to test the values of the cube and cylindrical compressive strength, splitting tensile strength, flexural tensile strength and the modulus of elasticity of the concrete. Tensile tests have been used to determine the average value of the yield strength of the structural steel, and composite action has been verified by push-out tests.

Characteristic values of the concrete strength and the corresponding mechanical characteristics necessary for the design of concrete components are provided in STN EN 1992-1 1, Part 3, for each strength class of concrete and in STN EN 206-1:2002. The Eurocode stipulates the procedures by which material characteristics are to be calculated for the compressive strength of concrete at particular ages [1, 2].

The application of the design values provided by the STN EN 1992-1-1 standard is adequate for the majority of cases. If the full potential of a concrete deck is to be exploited, it has been discovered that the assumed design values limit design possibilities. For instance, in the design of decks, bending is commonly the factor limiting their spans, and it is related to the average modulus of elasticity. If a higher value of the modulus of elasticity could be considered, the span of a concrete deck could be increased without increasing the thickness (depth) of the deck. Hence, using more precise material characteristics of concrete, a higher value can be determined than the assumed value of the characteristic in question [3, 4].

For the tests performed in the Laboratories Faculty of Civil Engineering at the Technical University of Košice, ready-mix concrete was delivered fresh from a nearby central mixing plant. It was necessary to verify the actual compressive strength of the concrete. Concrete samples used in the tests to specify the actual material characteristic of the site concrete were made along with the manufacture of experimental specimens to determine their bending resistance. All concrete samples (cubes, cylinders and blocks) intended for the specification of their strength characteristics complied with the requirements laid down by the STN EN 12390-1 standard [5, 6]. Standard cubes 150 mm wide and cylinders 300 mm high were used for the determination of

© The Author(s), under exclusive license to Springer Nature Switzerland AG 2021
V. Kvočák and D. Dubecký, *Research and Development of Deck Bridges*,
SpringerBriefs in Applied Sciences and Technology,
https://doi.org/10.1007/978-3-030-66925-6_4

Fig. 4.1 Concrete cubes for a cube crushing test in the testing machine and after failure

Table 4.1 Cube compressive strength of concrete at 28 days

Item no	Designation	Dimensions $d_1 \times d_2 \times d_3$ (mm)	Weight (kg)	Load (kN)	Cube compressive strength (MPa)
1	016/12/1/a N1-K1	149.2 × 149.2 × 150.0	7.487	652	29.14
2	016/12/1/b N1-K2	149.9 × 149.7 × 151.1	7.502	615	27.19
3	016/12/1/c N1-K3	150.0 × 149.9 × 152.9	7.554	640	27.92
Average: 28.08 MPa					

the values of the cube and cylindrical compressive strengths of the concrete. Before the samples were placed in a crushing device, all material residues were removed from their surface, and a thin layer of levelling finish worked into the surface to flatten it. A steel ring cup, as shown in Fig. 4.1, was attached centrically to the upper edge of the cube. No pads were placed in-between the cube and the plates of the cube testing machine. The load was applied perpendicularly to the concrete layering, evenly, without any impact, gradually increasing from 0.2 to 1.0 MPa/s. At the end of the test, the cubes failed in a usual manner, being crushed between the plates of the machine (Fig. 4.1).

Crushing tests took place within a period of $t = 28$ days, during which period static and fatigue tests took place as well. The results of the tests are provided in Tables 4.1, 4.2, 4.3 and 4.4.

A similar method was applied to test the cylinder compressive strength of the concrete, as shown in Fig. 4.2. The values of the cylinder compressive strength of concrete are used in Eurocode 2 [1] as a basis for most design calculations as in some situations these values are very close to the values of the amount of load at the concrete failure. For example, in a simple beam without shear or upper reinforcement, the value of the amount of load at the concrete failure approximates the value

Table 4.2 Cube compressive strength of concrete—testing the first beam

Item no	Designation	Dimensions $d_1 \times d_2 \times d_3$ (mm)	Weight (kg)	Load (kN)	Cube compressive strength (MPa)
1	024/12/1/a N1-K7	149.7 × 149.8 × 149.7	7.726	840	37.38
2	024/12/1/b N1-K8	149.5 × 149.5 × 149.6	7.636	870	38.88
3	024/12/1/c N1-K9	149.5 × 149.6 × 149.6	7.608	880	38.88
Average: 38.56 MPa					

Table 4.3 Cube compressive strength of the concrete—testing the second beam

Item no	Designation	Dimensions $d_1 \times d_2 \times d_3$ (mm)	Weight (kg)	Load (kN)	Cube compressive strength (MPa)
1	034/12/1/a N1-K10	149.8 × 149.9 × 150.1	7.79	870	38.60
2	034/12/1/b N1-K11	149.5 × 149.5 × 149.6	7.244	895	40.06
3	034/12/1/c N1-K12	149.5 × 149.7 × 151.4	7.605	900	40.35
Average: 39.72 MPa					

Table 4.4 Cube compressive strength of concrete – testing the third beam

Item no	Designation	Dimensions $d_1 \times d_2 \times d_3$ (mm)	Weight (kg)	Load (kN)	Cube compressive strength (MPa)
1	050/12/1/a N1-2-K1	149.6 × 149.5 × 149.6	7.642	895	40.03
2	050/12/1/b N1-2-K2	149.9 × 149.7 × 151.5	7.608	905	39.95
3	050/12/1/c N1-2-K3	149.4 × 149.4 × 149.9	7.634	915	40.86
Average: 40.28 MPa					

of the cylinder compressive strength of concrete, when taking into account the difference between the strength of concrete in cylindrical samples and the real structural member. On the other hand, in concrete members where the concrete is enclosed by reinforcement, such as in columns with shear reinforcement, the compressive stress the concrete can withstand before failure is remarkably higher. Eurocode makes appropriate allowances for that in design equations.

Fig. 4.2 Concrete cylinders for a cylinder crushing test in the testing machine and after failure

Table 4.5 Cylinder compressive strength of concrete at 28 days

Item no	Designation	Dimensions $d_1 \times L$ (mm)	Weight (kg)	Load (kN)	Cylinder compressive strength (MPa)
1	017/12/1/a N1-V1	147.8 × 302.8	11.19	438	25.44
2	017/12/1/b N1-V2	148.4 × 299.3	11.55	455	26.32
3	017/12/1/c N1-V3	148.0 × 300.6	11.43	440	25.57

Average: 25.78 MPa

The cylinder compressive strength of the concrete $f_{ck,cyl}$ (f_{ck}) was specified for cylinders 150 mm in diameter and 300 mm in height compliant with the STN EN 12 390–3 standard [7]. The results of the tests are displayed in Tables 4.5, 4.6, 4.7 and 4.8.

4.1 Flexural Tensile Strength Test of Concrete

Concrete blocks with the dimensions of 100 × 100 × 400 mm at 28 days were used for flexural tensile strength tests. The samples were subjected to a load that caused bending. The maximum load at which the sample failed determined the value of the flexural tensile test of the concrete.

Table 4.6 Cylinder compressive strength of concrete—testing the first beam

Item no	Designation	Dimensions $d_1 \times L$ (mm)	Weight (kg)	Load (kN)	Cylinder compressive strength (MPa)
1	025/12/1/a N1-V4	149.7 × 301.6	11.18	680	32.78
2	025/12/1/b N1-V5	149.8 × 300.4	10.91	685	31.55
3	025/12/1/c N1-V6	149.5 × 302.1	10.75	670	32.18

Average: 32.17 MPa

Table 4.7 Cylinder compressive strength of concrete—testing the second beam

Item no	Designation	Dimensions $d_1 \times L$ (mm)	Weight (kg)	Load (kN)	Cylinder compressive strength (MPa)
1	035/12/1/a N1-V7	149.4 × 301.2	10.70	590	33.26
2	035/12/1/b N1-V8	149.5 × 299.7	10.65	595	33.97
3	035/12/1/c N1-V9	149.5 × 302.7	10.85	590	33.66

Average: 33.63 MPa

Table 4.8 Cylinder compressive strength of concrete—testing the third beam

Item no	Designation	Dimensions $d_1 \times L$ (mm)	Weight (kg)	Load (kN)	Cylinder compressive strength (MPa)
1	051/12/1/a N1-2-V1	149.7 × 301.5	11.36	605	34.37
2	051/12/1/b N1-2-V2	149.5 × 303.0	11.53	625	35.60
3	051/12/1/c N1-2-V3	149.4 × 301.7	11.28	600	34.00

Average: 34.66 MPa

A compression testing machine, ADR ELE 2000, was used, consisting of two supporting cylinders and two upper loading cylinders. The upper cylinders were joined to an articulated cross-piece, able to move against each other, which distributed the load evenly on both sides. The distance between the supports was 300 mm, and the loading forces 100 mm apart from each other as shown in Fig. 4.3.

The speed of loading was held constant, without any changes or sudden impacts.

Fig. 4.3 Concrete blocks for a flexural tensile strength test of concrete

Table 4.9 Flexural tensile strength of concrete at 28 days

Item no	Designation	Dimensions $d_1 \times d_2 \times L$ (mm)	Weight (kg)	Load (kN)	Flexural tensile strength (MPa)
1	019/12/2/a N1-T1	100.1 × 100.2 × 400.2	9.224	11	3.30
2	019/12/2/b N1-T2	100.0 × 100.1 × 400.1	8.984	14	4.20
3	019/12/2/c N1-T3	100.0 × 100.2 × 400.1	9.156	12	3.60

Average: 3.70 MPa

The results of the test are displayed in Table 4.9.

The values of the flexural tensile strength of concrete are specified by the expression [8]:

$$f_{cf} = \frac{F \cdot l_0}{d_1 \cdot d_2^2} \qquad (4.1)$$

f_{cf} is the value of the flexural tensile strength of concrete in *MPa* and *F* is the maximum force reached when the failure of the concrete blocks occurred, in *N*.

4.2 Splitting Tensile Strength of Concrete

There are three "kinds" of concrete tensile strength: flexural tensile strength, splitting tensile strength, and axial tensile strength, where there is no standard test for the last one. It is essential to realise that there is no single strength value, representing all three "kinds" of concrete tensile strength for a specific concrete as they vary, depending on a method of testing, speed of increasing a load and the size of a test sample. Discrepancies in values can be explained by the principle of the so-called weak link, which assumes that the failure in tension will start in the weakest element, and once started, it is highly probable that it will quickly spread up all along the cross-section. Therefore, if a greater area is subjected to tension, it is likely that the strength of the "weak link" will be lower than that of a smaller area; thus, the measured tensile strength will be lower as well. The result in practice is that the values of the tensile strength measured during a flexural tensile strength test will be considerably higher than those obtained in a splitting tensile strength test. EN 1992-1-1 [1] defines the term or concept of *tensile strength* as the value of the highest stress acquired in uniaxial tension, for example, as the axial tensile strength.

Tests of the splitting tensile strength of concrete were performed according to the EN 12390-6 standard [5, 7]. A hydraulic press Matest3000 and half-cylinder supports were used, as shown in Fig. 4.4. A load was applied and held constant until the samples failed. The values of the splitting tensile strength were calculated from the maximum value of the amount of load applied until failure on three samples. The results are presented in Table 4.10.

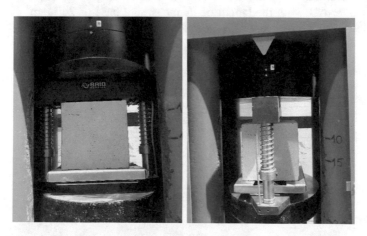

Fig. 4.4 A splitting tensile strength test of concrete

Table 4.10 Splitting tensile strength of concrete at 28 days

Item no	Designation	Dimensions $d_1 \times d_2 \times d_3$ (mm)	Weight (kg)	Load (kN)	Splitting tensile strength (MPa)
1	018/12/3/a N1-K4	149.4 × 149.4 × 149.8	7.314	76	1.80
2	018/12/3/b N1-K5	149.5 × 149.5 × 149.9	7.453	80	1.90
3	018/12/3/c N1-K6	149.4 × 149.4 × 150.0	7.461	78	1.85

Average: 1.85 MPa

4.3 Modulus of Elasticity of Concrete

Experimental specimens made of hardened concrete, $100 \times 100 \times 400$ mm in size, were used for the determination of the static modulus of elasticity for concrete in compression (Fig. 4.5a).

A hydraulic press Matest3000, compliant with the STN EN 12390-4 standard requirements (Fig. 4.5b), was used to load the specimens. The device can change loads over time and maintain a specific load at a desired level. For plotting the residual branch of a load-deformation graph (a stress–strain curve) for concrete, it is essential that the device can control and regulate the amount of load in correlation with the specimen deformation. Elongations were measured using a deformeter, EDA 250 Huggenberger AG, with the length of its measuring base of 250 mm. Extensions were detected at two opposite sides of the specimen.

(a)   **(b)**

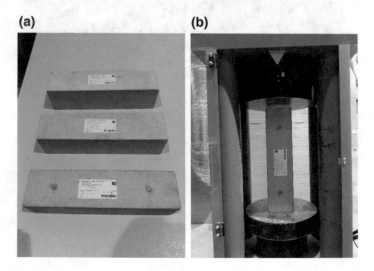

Fig. 4.5 a Prepared concrete blocks. **b** Test modulus of elasticity of concrete

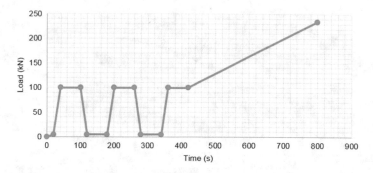

Fig. 4.6 Loading cycles over time

The procedure of the experimental specification of the static modulus of elasticity was conducted as follows: The experimental specimen was positioned in the centre of the test machine. A basic load/stress $\sigma_b = 0.5$ MPa was applied. The values of the forces and the deformations induced in the specimen were continuously measured and recorded. The intensity of the loading stress in the cross-section was controlled and uniformly increased over time, at a speed of 0.5 MPa/sec, until its value reached $\sigma_a = 10$ MPa, which was at 1/3 of the expected value of the compression strength, and this magnitude was kept constant for 60 s at the same load.

Subsequently, two more loading cycles with the σ_a and σ_b loads were completed. At each loading level, the constant loading stress was maintained for 60 s. It was necessary to monitor deformations during the whole experiment so that they did not differ from the average value by more than 20%.

Upon the completion of the last loading cycle and the 60 s waiting period, the load was increased in the experimental specimen in a specific manner until its rupture. A graph shows the process of loading the specimen in cycles in Fig. 4.6.

Average strains ε_a and ε_b were calculated from the individual measurements on both sides A and B of the specimen for each loading level with the loading stresses σ_a and σ_b (Fig. 4.7).

The static modulus of elasticity for concrete in compression, E_c (N/mm^2), is provided by the following equation [9, 10]:

$$E_c = \frac{\Delta\sigma}{\Delta\varepsilon} = \frac{\sigma_a - \sigma_b}{\varepsilon_a - \varepsilon_b} \tag{4.2}$$

σ_a is the upper loading stress in N/mm^2 (1/3 of the maximum magnitude of the f_c)
σ_b is the basic loading stress (0.5 N/mm^2)
ε_a is the average strain caused by the upper loading stress σ_a
ε_b is the average strain caused by the basic loading stress σ_b.

Testing the static modulus of elasticity of concrete has been quite rare until recently. It is still assumed that the values specified in the relevant standards are well applicable. Nonetheless, test results in the last few years have indicated that

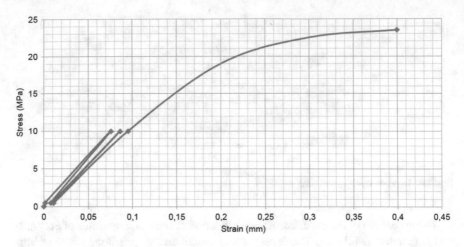

Fig. 4.7 Correlation between the stress and deformation for the determination of the static modulus of elasticity

Table 4.11 Modulus of elasticity of concrete in compression

Force (kN)	A	B	Absolute deformation (mm)	Reduced absolute deformation (mm)	Strain (μm/m)	σ_c (MPa)
0	−0.144	−0.160	−0.1520	0.0000	0.00	0
5	−0.144	−0.163	−0.1535	−0.0015	−6.00	0.5
100	−0.271	−0.184	−0.2275	−0.0755	−302.18	10
5	−0.147	−0.173	−0.1600	−0.0080	−32.02	0.5
100	−0.273	−0.203	−0.2380	−0.0860	−344.21	10
5	−0.147	−0.145	−0.1460	0.0060	24.01	0.5
100	−0.276	−0.188	−0.2320	−0.0800	−320.19	10
Resulting $E_{cm(t)}$= 33423.8 MPa						

the modulus of elasticity of concrete should be more seriously dealt with as early as at the stage of concrete design. The measurements completed in the experiments in question are presented in Table 4.11.

4.4 Tensile Test of Steel

Tensile tests of the steel used in the production of the beams were conducted by the Department of Material Science at the Metallurgical Faculty of the Technical University of Košice. Static tensile tests were performed using a breaking machine, ZWICK

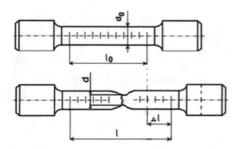

l_0 – the length of a test specimen
l – the length of the test specimen
at the moment of breakage
Δl – the tensile elongation of the
test specimen

Fig. 4.8 A steel specimen for a tensile test

387, and a traction dynamometer/load gauge, ZWICK WNo. 52497, following the STN EN 10002-1 specifications. The principle of the test rests in pulling apart a specimen of a standard shape and dimensions in the machine until breakage. A test specimen (as depicted in Fig. 4.8) was gripped and fixed into the jaws of the test machine so that the specimen axis was aligned with the axis of the force applied. Alignment of the test specimen in the test machine is critical to prevent exerting a bending force on it.

By the evaluation of the parameters measured in a static tensile test, the yield strength, tensile strength and ductility of steel were calculated. Test bars were designed in a manner that their initially measured length L_0 complied with the expression $L_0 = 10d_0$. To detect their ductility precisely, a notch was made on each bar to mark its initially measured length L_0 before the experiment. Then, the bar length was divided into ten parts with division marks (Fig. 4.8). The correlation between the tensile elongation of the specimen and the loading force was recorded on a graph during the test.

Three equivalent tensile tests were carried out for the given type of beam. Average values of the measured parameters were then calculated, which were regarded as nominal values and later used in the calculations of theoretical ultimate loads and the evaluations of the results. The values of the ultimate forces measured during the tests and the calculated yield strength, tensile strength of the steel and its ductility are provided in Table 4.12, while the stress–strain curves for the individual steel bars are displayed in Fig. 4.9.

Stress–strain diagrams of the individual specimens are displayed in Fig. 4.9.

Table 4.12 Results of tensile tests of steel

Item no	f_y (MPa)	$f_{y, ave}$ (MPa)	f_u (MPa)	$f_{u, ave}$ (MPa)	A (%)
1	317	315.3	397	396.0	41.7
2	313		397		41.7
3	316		394		40.0
Average value: 315.3 MPa					

Fig. 4.9 Stress–stress curves
for the three steel specimens

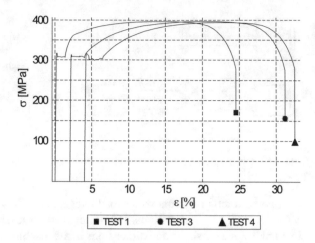

4.5 Push-Out Tests

Basic types of shear connectors, providing the interconnection between steel and concrete members used in composite beams, include, among others, also studs and strip connectors. Composite action ensured by studs, which are considered to be standard technological means produced with a high degree of automation, bears several adverse consequences. The main disadvantage of stud connectors is non-continuous force transfer between the composite elements within a structural member that causes the concentration of forces. The locations where the studs are welded are frequently sources of the initiation of cracks that propagate into the flanges of the main beams. From the point of fatigue failure, such structural detail is the most inappropriate in a composite beam, especially in bridges. Another drawback of stud shear connectors is their production cost and laboriousness, lengthening the construction period.

In most cases, the ultimate load-bearing capacity of a composite structural member is determined directly by the load-bearing capacity of the stud; therefore, it is impossible to raise the load-bearing capacity of a composite beam using higher-strength concrete. Unlike stud connectors, the majority of new-type shear connectors are not symmetrical regarding the axis perpendicular to the direction of shear flow. By their orientation, these shear connectors must be adjusted at least to the orientation of the prevailing shear flow. Strip shear connectors welded to the flange of a steel beam produce composite action only in a longitudinal direction; thus, they must be supplemented with elements preventing the uplift of a concrete deck. One of the first lying strips used in combination with studs was designed and applied in practice by Leonhardt in 1938. A standing perforated strip, first developed in Germany in 1985, is frequently used to ensure composite action both in vertical and in horizontal directions. Research into strip connectors has gradually progressed in other countries. The first results appeared in Canada in 1988 [11], and their complex analysis complemented with other results was published in 1992 [12]. Further research studies

were published in Australia in 1992 and Japan in 1994 [13], followed by the Czech
Republic in 1996 [14]. In Slovakia, the first investigations of shear strip connectors
were reported in 1996 [15].

Composite action provided by strip connectors rightfully deserves attention as,
when compared to the composite action provided by the traditional headed shear
studs, it seems more appropriate in terms of resistance and flexure. Moreover, strip
connectors exhibit better fatigue strength and durability and ensure continuous shear
force transfer to the flange of a beam. In comparison with stud connectors, the labour
intensity in the production of strip connectors is lower and the method of assembly
of a steel construction more straightforward, placing them once again before the
stud connectors. All the above properties predetermine this type of composite action
particularly for the application in mid-span composite bridges, which thus become
more economical, and materially and structurally more effective than pre-stressed
concrete bridges.

4.5.1 Perfobond and Strip Connectors

Composite action ensured utilising a perforated strip connector came to use in
Western Europe, where it has been protected by a patent since 1985. The advantage of
the employment of perforated/Perfobond or comb-like/Crestbond shear connectors
is that they are welded to a steel member directly in the process of manufacturing.
When being assembled on-site, before concreting a deck, reinforcement bars are
threaded through perforated holes in the strip.

The design resistance of a perforated strip connector depends on the geometrical
characteristics of the strip itself, and it must be determined for each type individually.

The findings of the research done at the Faculty of Civil Engineering at the Technical University of Košice suggest that the design resistance of a perforated strip
connector, as shown in Fig. 4.10, be calculated from the following expression [15]:

$$P_{Rd} = 1.57 \, f_{ck}/\gamma_v \text{ [kN]} \tag{4.3}$$

f_{ck} is the value of the cylinder strength of concrete and

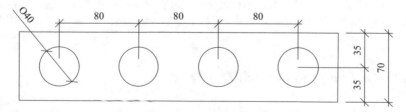

Fig. 4.10 Shape of a perforated strip connector used in a research study conducted at the Faculty
of Civil Engineering at the Technical University of Košice

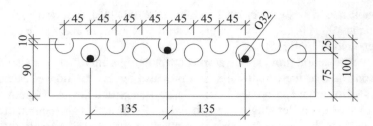

Fig. 4.11 Shape of a perforated strip connector used in the research study at ČVUT in Prague

γ_v is a statistically determined partial material factor, which equals 1.3 here.

The equation above (4.3) provides the design resistance for one hole in a perforated strip connector, according to Fig. 4.10.

Similar results were obtained in the research carried out at the Czech Technical University ČVUT in Prague. According to the information available, two types of strip connectors that differed in their height and hole diameter were tested there (Fig. 4.11).

The design resistance of a strip connector 50 mm high, with holes 32 mm in diameter, is given as:

$$P_{Rd} = (-68 + 12.1\, f_{ck} + 797\, A_{st})/\gamma_v \text{ [N/mm]} \tag{4.4}$$

and the design resistance of a strip connector 100 mm high, with holes 60 mm in diameter, is given as:

$$P_{Rd} = (273 + 14.1\, f_{ck} + 313\, A_{st})/\gamma_v \text{ [N/mm]} \tag{4.5}$$

A_{st} is the cross-sectional area of the reinforcement threaded through a hole in a strip connector [mm^2/mm], while the characteristic value of the yield strength of reinforcing steel $f_{sk} = 490$ MPa is taken into consideration,

γ_v is a statistically determined partial material factor, which equals 1.4 for a lower strip connector and 1.25 for a higher strip connector.

Based on the knowledge acquired on the value of slip capacity δ for the characteristic value of the shear resistance of a single connector P_{Rk}, perforated strip connectors cannot be classified as ductile shear connectors.

4.5.2 Performance of Push-Out Tests

This chapter deals with the composite action ensured by a closed U-shaped perforated strip. All preparations and the conduct of the experiment took place in the Laboratories of Faculty of Civil Engineering at the Technical University of Košice. In order

Fig. 4.12 Binding and placing reinforcement into the formwork

to simplify and shorten the preparation process, formwork for the experiment had been prepared from large-formatted foiled chipboard panels connected with screws. Steel sections were supplied and reinforcement bars bent by specialised companies. The ready-made steel reinforcement cages, shown in Fig. 4.12, were then inserted into the framework. The bottom edge of the steel box section was protected with polystyrene tiles so that the face of the steel element would not lean directly against the concrete block during the experiment.

The prepared specimens were then encased in fresh concrete. The concrete for the specimens was delivered from a central mixing plant so that the exact compound formula of the concrete mix could be maintained. To ensure the excellent quality of the specimens, all of them were adequately compacted and their surface levelled and smoothed (Fig. 4.13). After the concrete had hardened, the framework was removed. Finally, when the specimens reached the required strength, the composite action between the steel and concrete members was experimentally verified.

Fig. 4.13 Casting concrete in the framework

The specimens were manufactured in compliance with the STN EN 1994-1-1 Annex B standard specifications [16]. Composite action was ensured using a steel section made by welding a U-shaped 6-mm-thick steel plate to the bottom flange of a beam with protruding ends. Holes 50 mm in diameter were flame-cut in the webs at an axial distance of 100 mm. Transverse reinforcement bars 12 mm in diameter were threaded through every third hole in the beam. Identically, holes 50 mm in diameter were flame-cut at an axial distance of 100 mm in the top flange. The holes were arranged alternately, either in the webs or in flange of the beam, ensuring one hole in each cross-section.

The length of a rigid steel element providing composite action in a composite beam was 400 mm. Two steel box sections were fillet-welded with the help of steel plates, as demonstrated in Fig. 4.14. The double-box section created in this manner was able to transfer load evenly into both strip connectors. The steel sections were not coated with any protective paint so that the results of the experiment were not distorted.

The deformation mode was similar in all test specimens; however, failure of the box section in the first specimen occurred. As a result, all other specimens were then reinforced by adding steel plates on both sides of the section so that the deformation of the steel region not acting compositely was prevented. The specimens were subjected to a static load induced by a hydraulic press, INSTRON, generating a force up to 2,500 kN, as displayed in Fig. 4.16. Displacements between the steel sections and concrete blocks were measured using inductive and mechanical sensors, with a precision of 0.01 mm. Contact between the steel section of the specimen and the pressure area of the machine was provided through a steel bearing plate (Fig. 4.15).

Specimens of standard shape and dimensions compliant with STN EN 1994-1-1 were used in the push-out tests. The dimensions of the concrete block were 600 × 600 × 200 mm, and the length of the steel element (acting compositely) encased in concrete was 400 mm. The force applied was monitored and controlled during the experiment. The load was gradually increased with an increment of 10% of the estimated load-bearing capacity of the specimen to 40% of its previously calculated ultimate resistance. At this load level, the specimen was repeatedly unloaded and reloaded within a range of 10−40% of the estimated resistance.

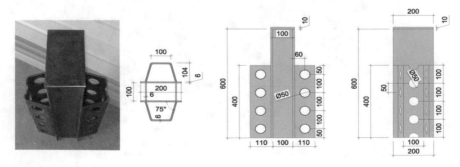

Fig. 4.14 A steel double-box section providing composite action in a composite beam

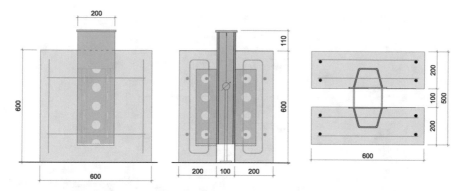

Fig. 4.15 Diagram showing the arrangement of a push-out test

Fig. 4.16 A hydraulic press and a specimen after the completion of a push-out test

Upon twenty-five unloading cycles, the test proceeded step by step until the design resistance of the specimen was reached. Apart from the amount of slip, the rate of its increase was determined at each loading stage. That was recorded one minute after the required loading level was reached and then, if necessary, every five minutes until the slip consolidated. The slip was regarded as steady when the rate of its increase dropped below 0.01 mm/min. The experiment was completed when the loading force became uncontrollable and impossible to be increased any longer.

Sound emissions were registered in the course of the experiment caused by cracking concrete. That was accompanied temporarily by a decrease in the force

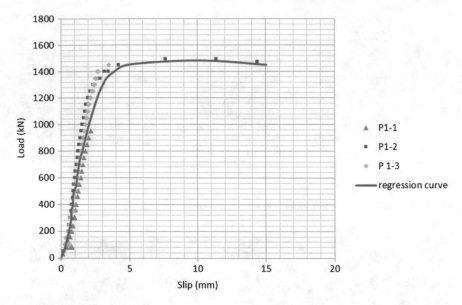

Fig. 4.17 Load-slip dependence

and a sudden increase in the slip by several thousandths of a millimetre. After a certain time, the force stabilised and returned to its original magnitude. The results of push-out tests are presented in Fig. 4.17.

References

1. STN EN 1992-1-1 Eurokód 2: Navrhovanie betónových konštrukcií, časť 1-1: Všeobecné pravidlá a pravidlá pre pozemné stavby
2. STN EN 206-1: 2002 Betón, časť 1: Špecifikácia, vlastnosti, výroba a zhoda
3. J. Bilčík, Ľ. Fillo, J. Halvoník, Betónové konštrukcie (Bratislava, 2005)
4. J. Brooks, Pružnosť, zmrašťovanie, dotvarovanie a teplotné pohyby. Advanced Concrete Technology. Vlastnosti betónu, published by John Newman and Ban Seng Choo (2003). ISBN 0 7506 5104 0
5. STN EN 12390-1/AC:2005: Skúšanie zatvrdnutého betónu, časť 1: Tvar, rozmery a iné požiadavky na skúšobné telesá
6. STN EN 12390-3: Skúšanie zatvrdnutého betónu, časť 4: Pevnosť v tlaku
7. STN EN 12390-2:2001: Skúšanie zatvrdnutého betónu, časť 2: Výroba a príprava skúšobných telies na skúšky
8. STN EN 12390-5: Skúšanie zatvrdnutého betónu, časť 5: Pevnosť v ťahu pri ohybe skúšobných telies
9. STN ISO 6784:1993 Betón. Stanovenie statického modelu pružnosti v tlaku
10. S. Unčík, P. Ševčík, Modul pružnosti betónu, edícia betón rácio (Trnava, 2008). ISBN 978-80-969182-3-2. https://www.betonracio.sk/betonracio
11. P.J. Antunes, Behavior of perfobond rib shear connector in composite beams. Thesis (B.Sc.) The University of Saskatchewan, Saskatoon, Canada (1988)

12. M.R. Veldana, M.U. Hosain, Behavior of perfobond rib shear connectors in composite beams. Push-out tests. Canad. J. Civil Eng. **19**. ISSN 0315-1468
13. A. Li, K. Cederwall, Push-out tests on studs in high-strength and normal-strength concrete. J. Constr. Steel Res. **36**(1), 15–29 (1996)
14. J. Studnička, J. Macháček, K. Peleška, Nové prvky spražení pro ocelobetónové konstrukce. Stavební obzor **5**(2), 42–45 (1996). ISSN 1210-4027
15. M. Rovňák, A. Ďuricová, K. Kundrát, Ľ. Naď, Spriahnuté oceľovo-betónové mosty (Košice, Elfa s. r. o., 2006). ISBN 80-8073-485-2
16. STN EN 1994-1-1 Eurokód 4: Navrhovanie spriahnutých oceľobetónových konštrukcií. Časť 1–1: Všeobecné pravidlá a pravidlá pre budovy

Chapter 5
Static Loading Tests of Composite Beams

Short-term tests on experimental beams were conducted using a breaking machine. As the beam specimens were laid on a solid base during the concreting and concrete hardening phases, the zero-loading state after the placement of the beam on the supporting plates of the breaking machine corresponded to the dead weight of the beam.

5.1 Measuring Devices of Static Loading Tests

The overall deflection and strains were observed in the steel and concrete regions of a specific composite beam during the experiments. The mid-span deflections and sagging of the specimens at their supports were measured using inductive sensors. The VSL-HBM L 200 W-type inductive sensors were employed to measure the mid-span deflections in the test specimens, whereas the VSL-HBM L 100 W-type inductive sensors were employed to verify the measurements of possible deformations in the test structure, as shown in Fig. 5.1.

All output data was recorded using two Spider 8 data buses connected to a portable PC. The deflections and strains (relative deformations) recorded by the strain gauges attached to the beam on the rigid steel reinforcement were analysed and evaluated using the Catman software environment.

Mid-span deformations in the compressed concrete region at the points of concreted targets were measured using the EDA 250 Huggenberger AG deformeter (with a 250 mm measuring base). All values were noted down by hand (Fig. 5.2).

Strains in the steel beams were monitored using the HBM 1-LY11-6/120 strain gauges. These were bonded on the beams with the aid of a transportable case, containing a set of two-component epoxy glues, strain gauges and all accessories needed for bonding (Fig. 5.3).

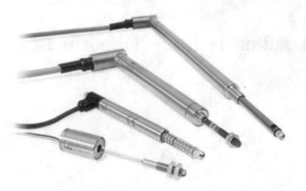

Fig. 5.1 Inductive sensors for the measurement of deflections

Fig. 5.2 Measurements of deformations in the compressed concrete region using the Huggenberger AG-EDA 250 deformeter

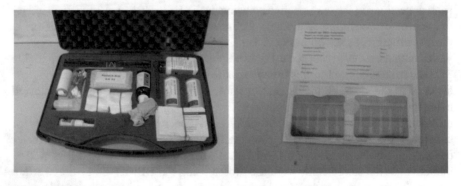

Fig. 5.3 A set for bonding strain gauges and the HBM strain gauges in the case

Fig. 5.4 A manual loading press

A test specimen was loaded with two hydraulic presses of the ENERPAC system, while the values of the compression forces induced in the system were controlled using a calibrated manometer within a range between 0 and 700 bars (Fig. 5.4).

5.2 The Experimental Set-up and the Placement of Strain Gauges

A test specimen was loaded by two vertical forces, applied at a distance of 2000 mm from both edges, the axial distance between the forces being 1800 mm and a free end overhanging the support by 100 mm. It was loaded utilising two symmetrically positioned hydraulic presses (max. 160 kN) so that pure bending occurred in the section between the presses, as shown in Fig. 5.5.

The zero-loading state corresponded to the dead weight of the beam. The moment caused by the dead weight was $M_g = 27.33$ kNm. The loading pressure in the hydraulic machine increased incrementally by 5 kN. The load on the beam was reduced/released twice: first, at the load level of 60–15 kN in each hydraulic press; second, at the load level of 80 kN to 15 kN again. With each loading step, after the deformations stabilised, deflection and strain measurements were taken, using inductive sensors and strain gauges, respectively. The strain gauges were attached to the encased steel beam at the most bent locations. The inductive sensors measured the mid-span and end-support deflections (Fig. 5.6).

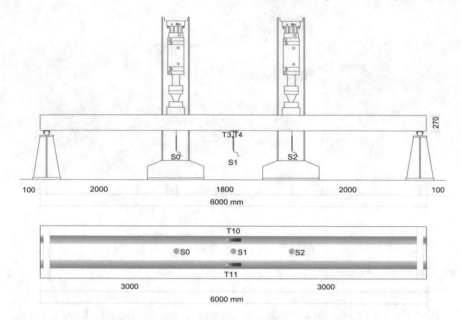

Fig. 5.5 Experimental set-up of a composite beam and the placement of strain gauges

Fig. 5.6 A specimen at the beginning and end of the experiment

After the load exceeded the tensile strength of concrete, hair-like cracks began to form and branched in the concrete that had been subjected to tension. These cracks further opened, propagated and enlarged to reach a length of approximately 200 mm, which was the estimated position of the plastic neutral axis in the beam.

The tests were completed when the load being transmitted by the specimen could not be increased any longer, and the deflection grew significantly even without adding any more load.

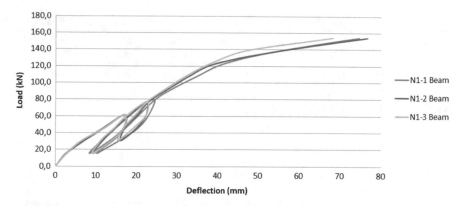

Fig. 5.7 Correlation between the deflection and the amount of load applied to a beam

Strains in the steel were measured and recorded with strain gauges positioned at the locations most subjected to bending and around the holes in the steel sections. Inductive sensors measured mid-span and end-support deflections. The correlation between the maximum deflection and the load applied is shown in Fig. 5.7.

5.2.1   Gradual Increase in the Compression Strength of Concrete Over Time

Table 5.1 shows the maximum forces F_{exp} with which the specimens were loaded by the presses at the end of the experiments along with the moments M_{exp} corresponding to the maximum load and the resulting experimentally determined average value of the ultimate resistance moment $M_{exp, ave}$. The resulting value of the resistance moment is compared to the numerically determined value in percentage.

Different values of the cylinder compressive strength of concrete were applied in the numerical calculations, the fact which was explained in more detail in Sect. 5.1. A time interval between the specification of the strengths at 28 days and at the time of the first N1-1 beam test was ten days, and the strength increased by 6.39 MPa. At the time of the second N1-2 beam test, eleven days later, the strength increased only by 0.74 MPa, and during the third N1-3 beam test, another eleven days after the second test, the strength rose by 0.72 MPa. It was assumed that additives and admixtures were used as the concreting process took place at low temperatures. Margins in the values of the resistance moment in the specimens under observation to those theoretically and numerically calculated amounted to 5.62, 6.16 and 7.68% for the respective beams, which equals 6.48% on average.

Table 5.1 Results of a static loading test and their comparison with analytical calculations

Specimen	F_{exp} (kN)	M_{exp} (kNm)	M_{theor} (kNm)	Difference %	$M_{exp, ave}$ (kNm)	$M_{theor, ave}$ (kNm)	Difference %
N1-1	154.0	335.33	317.48	+5.62	339.37	318.69	+6.48
N1-2	155.5	338.33	318.71	+6.16			
N1-3	158.5	344.44	319.88	+7.68			

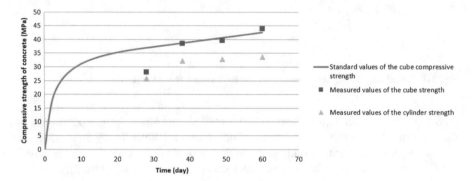

Fig. 5.8 A gradual increase in the compression strength of concrete over time

The following graph in Fig. 5.8 presents a gradual increase in the concrete strength of concrete over time and the tests performed at the same time.

5.3   Cracking in Reinforced Concrete and the Width of Cracks

When the stresses exceeded the tensile strength of concrete at some point, cracks arose in the concrete.

Provided that the high compression strength of concrete is to be taken as an advantage in a reinforced concrete structure, cracking is expected in the compressed reinforced concrete, and tension is assumed to be borne by the reinforcement. Although cracks can deteriorate the operational properties of the structure, when such crack widths and depths are limited to acceptable values, they should be of no concern. Crack patterns depend on the method of loading, dimensions of the element and the reinforcement. In the static loading test, bending cracks occurred at the bottom edge of the beam.

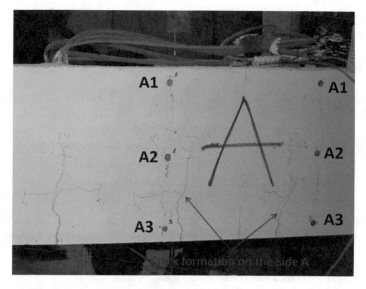

Fig. 5.9 Longitudinal strain monitoring points on the beam

During the static loading test, cracks arising in a composite beam were monitored; their widths and depths measured at all loading stages. The composite steel-concrete beam was divided into two sides, A and B. The initial loading stage was at a zero load. The first cracks appeared at the fourth stage of loading, with a force of 46.16 kN. By a gradual addition of load, the cracks widened and deepened. The load on the beam was released several times during the test; however, the cracked concrete would not allow the beam to return to its initial position. The ultimate crack width was reached at the twenty-first loading stage when the force was 123 kN. In total, the beam underwent 24 loading cycles, and the final amount of load applied was 155.4 kN.

Gradual propagation and penetration of cracks in the concrete are shown in Fig. 5.9. The cracks were continuously marked for all individual loading stages during the experiment as well as their opening, widening and deepening towards the neutral axis of the beam.

Figure 5.10 shows the correlation between the crack width and the amount of load applied. Slight deviations in the diagram were due to the different modes of crack opening, where only a single crack in the middle of the beam was accounted for, and the others arising in the tensile concrete region were neglected.

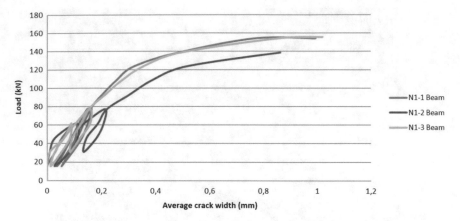

Fig. 5.10   Correlation between the crack width and the amount of load applied to a beam

5.4   Longitudinal Deformations and Strains

Three points were marked on the N1 beam, one in the middle and two more on both sides at a 250-mm distance. Elongations of the beam during the experiment were measured at these points using the Huggenberger extensometer. The measurements of the elongations on the beam took place at specific loading cycles. The values regarding the initial unloaded beam were recorded during the first cycle. Loading and unloading were then changed cyclically until the ultimate load was reached, followed by the final unloading step at the end of the test. At the same time, elongations were measured at all stages using the Huggenberger AG-EDA 250 deformeter. The measured values of these longitudinal deformations and strains in the specimen beam are provided in Table 5.2.

The correlation between the elongations and the load applied at the individual loading cycles is represented by the graphs in Figs. 5.11 and 5.12. They show the breaking point at which the ultimate fatigue limit state was exceeded, and the load culminated at the ultimate limit as well, followed by the final stage of unloading the beam (Figs. 5.13, 5.14, 5.15, 5.16 and 5.17).

As has been proven by the experiment, the neutral axis of the cross-section tends to move towards the upper compressed fibres in the composite member, which supports the underlying research assumptions.

Figure 5.18 clearly illustrates the breaking point at which the ultimate fatigue limit state was exceeded in the concrete zone of the beam. Moreover, the periodicity of the gradual load increase and decrease in the composite member can be seen there.

After loading the beam with a 60 kN force, and then an 80 kN force, the load was released, but despite the removal of the external load, the beam remained permanently deformed, with a severe deformation caused by the cracking of the concrete deck.

Table 5.2 Values of longitudinal deformations measured in the N 1-2 beam

LS	Pressure (bar)	Load (kN)	A1-1	A2-2	A3-3	B1-1	B2-2	B3-3
1	**0**	**0**	**−0,282**	**−0,132**	**0,401**	**−0,147**	**−0,421**	**0,23**
2	10	15,386	−0,305	−0,135	0,428	−0,206	−0,44	0,138
3	20	30,772	−0,333	−0,119	0,482	−0,231	−0,435	0,178
1. Crack formation on the side A and side B (LS 4)								
4	30	46,158	−0,36	−0,108	0,533	−0,271	−0,41	0,226
5	40	61,544	−0,388	−0,094	0,595	−0,303	−0,42	0,272
6	30	46,158	−0,386	−0,095	0,595	−0,317	−0,42	0,375
7	20	30,772	−0,366	−0,1	0,565	−0,291	−0,417	0,31
8	10	15,386	−0,334	−0,108	0,51	−0,252	−0,37	0,195
9	20	30,772	−0,355	−0,103	0,538	−0,269	−0,427	0,295
10	30	46,158	−0,374	−0,098	0,57	−0,298	−0,372	0,363
11	40	61,544	−0,394	−0,096	0,619	−0,319	−0,367	0,279
12	50	76,93	−0,416	−0,077	0,695	−0,328	−0,385	0,34
13	40	61,544	−0,418	−0,078	0,69	−0,339	−0,335	0,33
14	30	46,158	−0,404	−0,079	0,664	−0,328	−0,39	0,309
15	20	30,772	−0,376	−0,09	0,617	−0,301	−0,412	0,355
16	30	46,158	−0,381	−0,087	0,628	−0,304	−0,329	0,286
17	40	61,544	−0,406	−0,08	0,669	−0,302	−0,377	0,35
18	50	76,93	−0,423	−0,074	0,702	−0,355	−0,376	0,322
19	60	92,316	−0,449	−0,058	0,784	−0,384	−0,342	0,403
20	70	107,702	−0,48	−0,036	0,866	−0,413	−0,37	0,548
Exceeded limit state crack (LS 21)								
21	80	123,088	−0,521	−0,002	0,993	−0,442	−0,351	0,633
22	90	138,474	−0,608	0,084	1,346	−0,533	−0,267	0,877
23	100	153,86	−0,702	0,218	1,822	−0,645	−0,124	1,188
Max load								
24	101	155,3986	−0,763	0,32	2,156	−0,704	−0,051	1,482
Lightweight (LS 25−27)								
25	70	107,702	−0,763	0,315	2,141	−0,663	−0,039	1,445
26	40	61,544	−0,687	0,284	1,959	−0,61	−0,02	1,322
27	0	0	−0,546	0,245	1,656	−0,783	−0,02	1,226

From the viewpoint of the static loading test, four loading states at which significant changes in the beam occurred were most decisive. These are presented in Table 5.3.

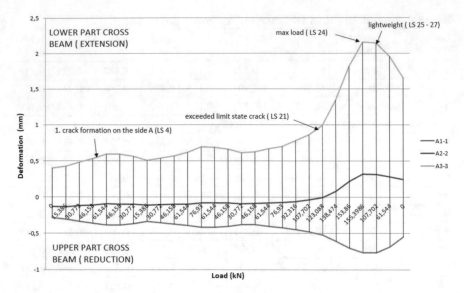

Fig. 5.11 Correlation between the longitudinal deformation and the amount of load applied to Side A of the beam

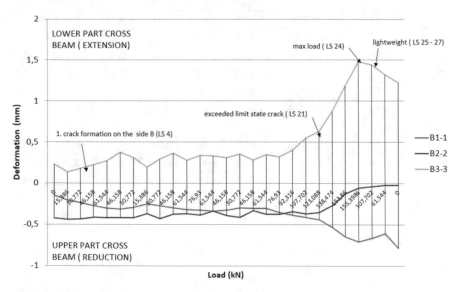

Fig. 5.12 Correlation between the longitudinal deformation and the amount of load applied to Side B of the beam

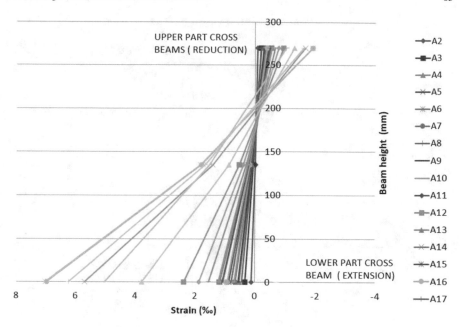

Fig. 5.13 Development of strains along the height of the beam—Side A

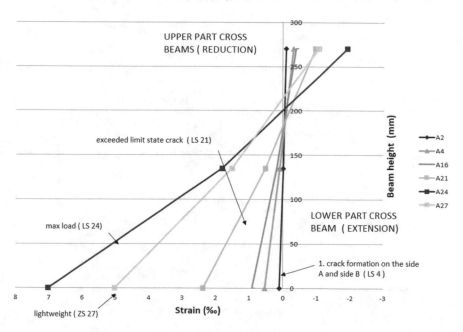

Fig. 5.14 Development of strains along the height of the beam—Side A—for some selected loading states

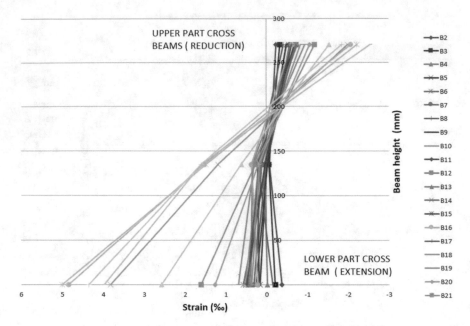

Fig. 5.15 Development of strains along the height of the beam—Side B

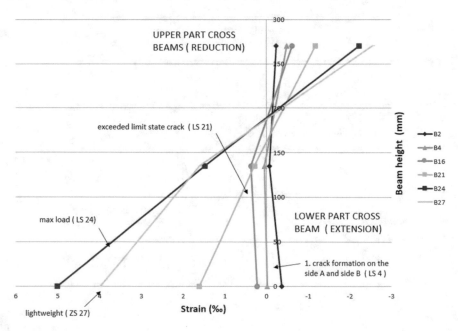

Fig. 5.16 Development of strains along the height of the beam—Side B—for some selected loading states

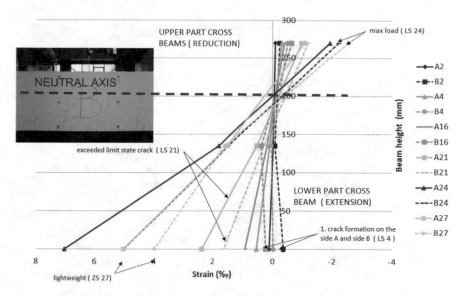

Fig. 5.17 Development of strains along the height of the beam—Sides A and B—for some selected loading states

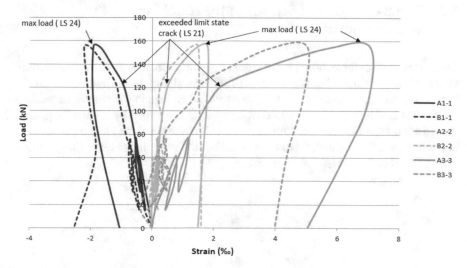

Fig. 5.18 Correlation between the strains and loads applied during the loading process

Table 5.3 Values of strains dependent on the force applied on the beam

Loading state	Monitoring point	Strains ε		
		Upper part 1-1 (‰)	Middle part 2-2 (‰)	Bottom part 3-3 (‰)
4	The first crack on both Side A and Side B on the beam F = 46.158 kN			
	Side A	$\varepsilon = -0.312$	$\varepsilon = 0.096$	$\varepsilon = 0.528$
	Side B	$\varepsilon = -0.496$	$\varepsilon = 0.044$	$\varepsilon = -0.016$
16	Another crack on Side A of the beam F = 46.158 kN			
	Side A	$\varepsilon = -0.396$	$\varepsilon = 0.180$	$\varepsilon = 0.908$
	Side B	$\varepsilon = -0.628$	$\varepsilon = 0.368$	$\varepsilon = 0.224$
21	The ultimate fatigue limit state F = 123.088 kN			
	Side A	$\varepsilon = -0.956$	$\varepsilon = 0.520$	$\varepsilon = 2.368$
	Side B	$\varepsilon = -1.180$	$\varepsilon = 0.280$	$\varepsilon = 1.612$
24	The ultimate load limitF = 155.399 kN			
	Side A	$\varepsilon = -1.924$	$\varepsilon = 1.808$	$\varepsilon = 7.020$
	Side B	$\varepsilon = -2.228$	$\varepsilon = 1.480$	$\varepsilon = 5.008$

5.5   The Evaluation of Strains in the Rigid Steel Reinforcement

Strains (relative deformations) in the rigid steel reinforcement were measured using strain gauges positioned at some previously selected and designated points on the beam. The measurement results point to a gradual change in the strains, depending on various load levels. Figure 5.19 shows the strains recorded by T3 and T4 gauges placed at the end of the steel beam and on the side of the steel box section between the holes. Besides, it is evident from the graph how the vertical stresses (T3) gradually redistributed to the horizontal ones (T4).

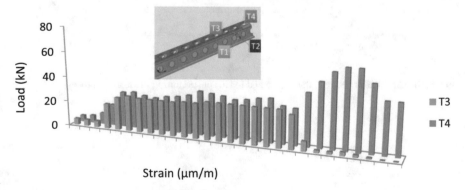

Fig. 5.19 A graph showing the load-strain correlation in the steel beam at T3 and T4 gauge points

Figure 5.20 represents the resulting strains measured using the strain gauge points T1 and T2 located at the end of the steel beam and the bottom edge of the steel box section web. As shown in the following graph, the vertical strains, due to the local deformations around the holes, gradually changed from tension to compression.

Finally, Fig. 5.21 represents the resulting strains in the mid-span of the steel beam. Although the rigid steel reinforcement was placed in the tension region of the composite beam, local compression deformations occurred between the holes (T8). Besides, it is evident from the graph below that the strain in the web on the side is remarkably lower than that in the top flange of the steel box section. Permanent plastic deformations started to arise in the section as the load exceeded 130 kN.

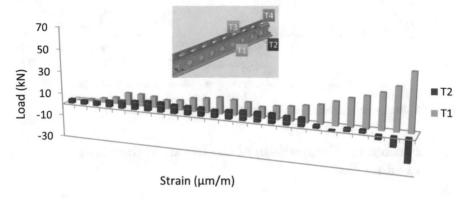

Fig. 5.20   A graph showing the load-strain correlation in the steel beam at T1 and T2 gauge points

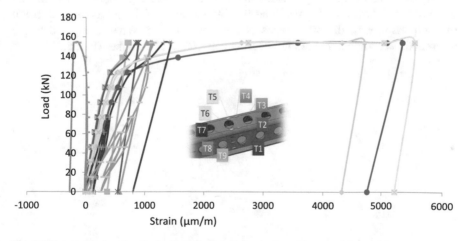

Fig. 5.21   A graph showing the load-strain correlation in the mid-span of the steel beam

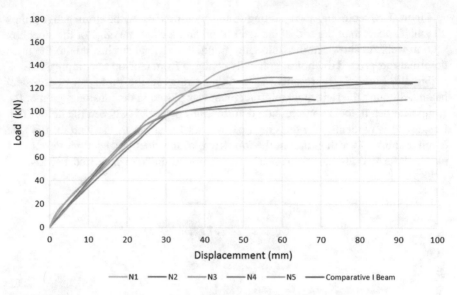

Fig. 5.22 Parametric comparison of the N1–N5 beams under static loads

5.6   Parametric Comparison of the Beams at the Static Loading Test

The following graph indicates the parametric comparison of the individual N1–N5 types of the beam using various methods of providing composite action. The graph clearly shows the highest load-bearing capacity of the N1 composite beam in comparison with the other types of composite action. When compared to the reference I-section beam, the load-bearing capacity is 24% higher (Fig. 5.22).

Chapter 6
Long-Term Tests of Composite Beams

For the purpose of determining rheological properties of the materials used, ten experimental beams were earmarked for long-term measurements. Two beams of each type (N1–N5) were subjected to long-term loading, lasting a minimum of 420 days. The magnitude of a compression force evenly applied on the beam was measured during the experiment, while the corresponding values of total and relative deformations were recorded for each loading cycle Fig. 6.1

The beams were placed on their sides during the long-term tests. There were always two beams making a pair/set that was supported and loaded simultaneously. They were turned with their top flanges facing each other, creating a 50 mm spacing between them where a pneumatic pillow was inserted. The support structure consisted of frames that compressed the beams against each other at the place of their theoretical supports.

6.1 Measuring Devices of Long-Term Tests

A U-shaped frame was made to secure the beam supports, with a stabilising threaded rod placed in the upper part to close the frame (Fig. 6.2). After the placement of the beams into the frame and insertion of the rubber pillow, the rod was tightened up. Then, the specimens were ready for incremental continuous loading with compressed air.

The specimens were loaded continuously using a rubber air pillow (Fig. 6.3) where compressed air was forced in.

A single-plunger oil compressor, Scheppach HC 50, with a maximum operating pressure of 8.9 bars (Fig. 6.4) was used to generate compressed air to blow into the pillows through quick-acting coupling plugs, fittings and pressure hoses.

Pressure in the rubber pillows was regulated using a manometric pressure controller, as shown in Fig. 6.5. Each air pillow was controlled separately so as

© The Author(s), under exclusive license to Springer Nature Switzerland AG 2021
V. Kvočák and D. Dubecký, *Research and Development of Deck Bridges*,
SpringerBriefs in Applied Sciences and Technology,
https://doi.org/10.1007/978-3-030-66925-6_6

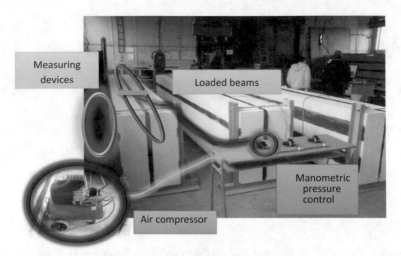

Fig. 6.1 A schematic view of the measurement system

Fig. 6.2 Load-bearing steel
frames serving as supports
during long-term tests

to provide an incremental increase of load. The magnitude of pressure applied was
monitored through a digital manometer Würth (Fig. 6.6).

Deflections were measured using the VSL-HBM L 200 W-type inductive sensors
(Fig. 6.7). The specimens were laid on the side to eliminate the deflections caused
by its dead weight. Relative deflections of the specimen against the solid floor were
observed and continually recorded for 420 days.

Fig. 6.3 A rubber air pillow

Fig. 6.4 Oil compressor
Scheppach HC 50

6.2 The Loading Procedure of Long-Term Tests

The load was activated by the pressure of a pneumatic pillow that was applied uniformly over the overall area of both panels. The constant pressure in the pillow was maintained with an air compressor interconnected with manometers though air-hose valves, while a different pressure could be set for each pillow.

The specimens were loaded by small incremental advances of 5 kPa to reach the final pressure of 30 kPa. The long-term compression that the specimens were exposed to corresponded to as much as 40% of their bending resistance. After 200 days, the load was raised to 35 kPa, and another increase in the load was made at 420 days, to 40 kPa. As was said above, the deflections were measured with inductive sensors, always in the mid-span and at the end support of the beam.

ASTM C 512–02 is the standard method that covers the determination of the time-dependent effects of the overall creep in concrete (the *basic creep* of concrete due to ageing plus *shrinkage* due to the drying of concrete). At the time of the load application, the load applied should not induce compression stress higher than 40% of

Fig. 6.5 A manometric pressure controller to control the load

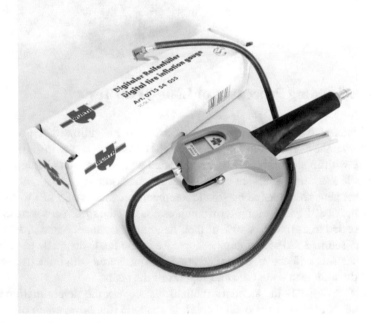

Fig. 6.6 A digital manometer Würth

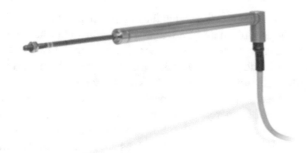

Fig. 6.7 Inductive sensors to measure deflections

Fig. 6.8 The loaded beams in the U-shaped steel frame

the compression strength. Readings were obtained immediately after the application of the load, and then after two and six hours. Consequently, readings were taken regularly at specified intervals for a year from the first loading.

The given method also stipulates how to calculate the rate of creep. According to Brooks, the equipment for the creep test under ASTM C 512 is enormous and expensive; therefore, as he claims, researchers tend to use smaller and cheaper devices. There is no standard European test included in the EN 12,390 testing standards; nonetheless, a method for testing materials intended for repairs of concrete structures is under development. This method is defined and annexed to the EN 13,584–2 standard.

Generally, when performing a creep test, the following should be taken into account: the rate of creep in the concrete subjected to sustained stress will be higher during its drying. Creep without drying (at constant total water content) is called *basic creep,* and the additional creep caused by drying is called *shrinkage.* For this reason, creep tests should be performed in the environment which closely resembles the conditions or the environment of a real construction site (where a real concrete structure works). Moreover, the effect of creep depends on the value of the concrete strength at the age of loading: the higher the strength at the time of loading, the lower the stress-strength ratio and the creep. Even with the same stress-strength ratio, the creep is lower if the load is applied on a concrete member with older age. Therefore, it is crucial to take the age of concrete at the time of loading into account for every creep test procedure. Creep increases in time; however, assumingly, it becomes constant after approximately 30 years. Creep attains values around 50% of the ultimate creep over the first two or three months, and the next 90% over the following two or three years. The longer the test, the more accurate the forecast for the long-term creep.

Gilbert explored mathematical formulations for the dependence of a creep coefficient on time and concluded that the equations forecasting the ultimate creep based on the data collected at 28 days were unreliable, and as a result, he advised longer creep tests. High temperatures also materially affect creep. Nevertheless, within the range of usual temperatures which structures are exposed to, the impact of temperature is relatively marginal in comparison with a humidity factor. For normal interior conditions where certain design specifications are required, standard requirements applicable in compliance with ASTM C 512–02 are enforced.

6.3   Long-Term Loading Tests Results

Long-term tests took place over 420 days. The beams were subjected to a continuous constant load. The effect of the environment was partially eliminated since the beams were installed inside a laboratory hall. The humidity ranged between 35 and 60%, and the temperatures fluctuated around +5 °C in winter and around 28 °C in summer. All long-term measurement results are shown in Fig. 6.9. The graph represents the gradual creep in the beam under continuous load, applied onto the beam using an air pillow (as described above). Deflections rose faster at the beginning, stabilised after a while, and continued to increase with the increasing load, yet very slightly. The tests proceeded with more load, and after another stabilisation, the load was increased for the third time. Time-and-load-dependent rheological changes were observed in the beam.

The long-term tests proved that the impact of rheological changes in concrete is significant at the beginning of the loading process. As could be drawn from the experiments, with a sustained constant load, these changes played a role in the first days after the loading. As much as 41% of the deflection took effect in the first week after the load application. When compared to Eurocode 2, the actual deflections throughout 420 days were only about 4% lower than the standard calculated values. Over a shorter period, the differences were even smaller.

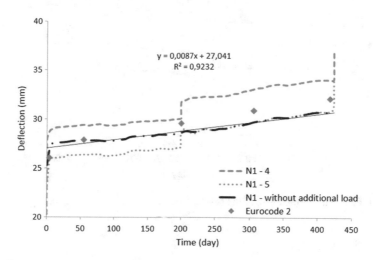

Fig. 6.9 A graph showing time-dependent deflections

6.4   Parametric Comparison of the Beams Under the Long-Term Loading

Long-term loading tests were performed on a total of five pairs of composite steel and concrete beams. Each pair came as one specific type of providing composite action between the beams, as was explained in more detail in Sect. 4.14.1. The N2–N5 specimens were loaded incrementally, with an addition of 5 kPa of load at each step until the value of 25 kPa was reached. The N1-type beam, with a higher load-carrying capacity, was loaded until a maximum of 30 kPa was generated. The compression to which all the specimens were subjected to for a long term corresponded to approximately 40% of their bending resistance. Deflections were measured separately in each beam within a specific pair of beams. The resulting average values are plotted in a graph in Fig. 6.10.

The graph represents the time-dependent deflections in the individual N1–N5 composite beams, also showing the gradual creep in the beams under continuous load applied onto the beams using an air pillow. Time-and-load-dependent rheological changes were observed in the beams.

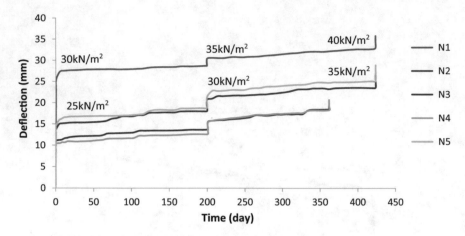

Fig. 6.10 Parametric comparison of the N1–N5 beams under the long-term loading

Chapter 7
Fatigue Tests of Composite Beams

The structural breakdowns that sporadically occurred in the past were attributed to both the wrong design and construction technology, but also to randomly generated extreme loads and impacts not accounted for in the structural design. *Fatigue* as a new term in the structural field was coined for the first time by Jean-Victor Poncelet, a French mathematician and engineer, around 1812–1814. Later, one of the most recognised founders of the fatigue theory and experimental determination of fatigue properties was August Wöhler, who first conducted tests where specimens were subjected to repeated cycles of applied loadings. His work marks the first systematic investigation of the so-called *Wöhler curves*, also known as S–N curves, characterising the fatigue behaviour of materials, and also the derivation of the so-called *fatigue limit*, which was then applied in calculations to define a reduced static strength to avoid potential failures of structures in operation [1, 2].

7.1 The Basic Fatigue Stress Curve

A *Wöhler curve*—also known as a stress endurance curve—is used for the verification of fatigue properties of materials, structural members or structures as such. It provides the correlation between the stress amplitude σ_a and the number of cycles N until fracture. The results for plotting the curve are obtained in fatigue tests with the so-called soft loading [3, 4].

Fatigue tests with soft loading cycles are historically the oldest and are performed on a hydraulic pulser (vibrator) with force control. The test results are plotted in the form of σ_a-N curves. These Wöhler curves can be constructed for various mean values of stress σ_m, which affect their shapes. In Fig. 7.1, there is a Wöhler curve plotted for an alternating symmetric cycle [5].

© The Author(s), under exclusive license to Springer Nature Switzerland AG 2021
V. Kvočák and D. Dubecký, *Research and Development of Deck Bridges*,
SpringerBriefs in Applied Sciences and Technology,
https://doi.org/10.1007/978-3-030-66923-6_7

Fig. 7.1 Individual fatigue regions as displayed on a Wöhler curve 1. The region of quasi-static fracture, which takes place after several tens of loading cycles. 2. The low-cycle fatigue region, characterised by cyclic stresses above the static yield strength. 3. The high-cycle fatigue region, characterised by cyclic stresses below the static yield strength

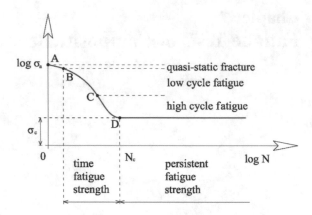

7.2　Measuring Devices of Fatigue Tests

Fatigue tests were conducted on a breaking machine in the Laboratory of Excellent Research Faculty of Civil Engineering at the Technical University of Košice. A pulsating load was generated using the Inova hydrodynamic pulser with hydraulic motors up to 650 kN (Fig. 7.2).

Strains in the steel sections were measured using Spider 8, a PC-based strain measurement system, as shown in Fig. 7.3.

Fig. 7.2　A hydrodynamic pulser with a hydraulic motor

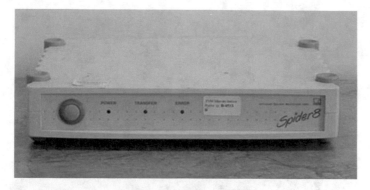

Fig. 7.3 A PC-based strain measurement system Spider 8

Fig. 7.4 A VSL-HBM L
200 W inductive sensor with
a 300 Hz sampling frequency

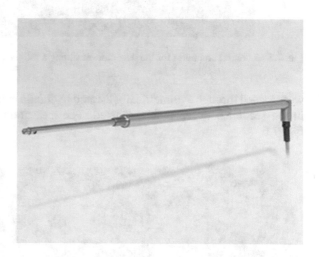

Deflections in the experimental beam were controlled by the VSL-HBM L 200 W inductive sensors, as shown in Fig. 7.4. The results were continuously recorded on a central computer with a 300 Hz sampling frequency (Fig. 7.5).

7.3 The Loading Procedure of Fatigue Tests

The goal of fatigue tests was to determine the fatigue strength of the designed specimens at failure under a cycling load. A bending fatigue test was selected where the load on a test specimen was introduced through two vertical forces 2,000 mm from the edges of the beam, while the spacing between the forces was 1,800 mm and the free end/cantilever protruded over the support of the beam by 100 mm.

Fatigue tests on the experimental beams took place on the same breaking machine as the one that was used for short-term tests under a quasi-static load. The specimens

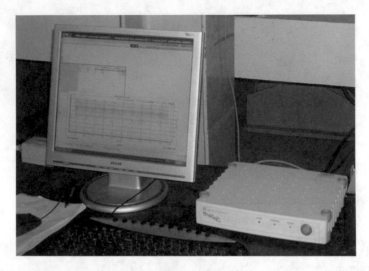

Fig. 7.5   A central computer for the data storage during a fatigue test

were loaded through symmetrically located hydraulic cylinders so that pure bending in the section between them occurred. The measurement set-up is evident from Fig. 7.6.

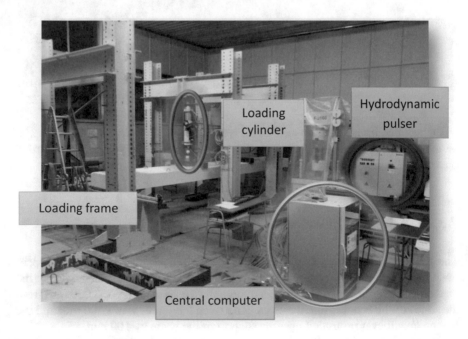

Fig. 7.6   Measurement set-up for a fatigue test

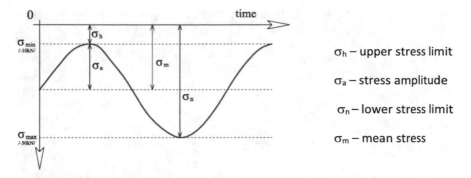

Fig. 7.7 A graph showing the fatigue loading cycle

The tests were performed by applying a cyclic load on three beams. The first one was subjected to the variable load with a loading cycle per cylinder ranging from − 10 to −90 kN, corresponding to approximately 66% of the theoretical load-bearing capacity of the beam and the fatigue stress range $\Delta\sigma_1 = 350$ MPa. Another beam was loaded within a load range between −10 and −50 kN, approximating 33% of its theoretical load-bearing capacity and the fatigue stress range $\Delta\sigma_2 = 175$ MPa. Finally, the last beam was subjected to a cyclic load between −10 kN and −30 per cylinder, corresponding to 16% of its theoretical load-bearing capacity and the fatigue stress range $\Delta\sigma_3 = 87.5$ MPa. The loading cycle, i.e. the stresses fluctuating between the upper and lower stress levels, is shown in Fig. 7.7. Failure in the beams occurred due to the initiation of a fatigue crack in the encased steel section, approximately at the point of load application. The results of the fatigue test on the N1 beam are presented in Fig. 7.10 (Figs. 7.8 and 7.9).

7.4   Fatigue Assessment and the Measurement Results

Nominal stress, modified nominal stress or effective stress ranges due to cyclic loads $\psi_1\,Q_k$ must not exceed:
    for the normal stress ranges

$$\Delta\sigma \leq 1.5\,f_y \quad 350 \leq 1.5 \cdot 317 \quad 350\,\text{MPa} \leq 475.5\,\text{MPa} \qquad (7.1)$$

for the shear stress ranges

$$\Delta\tau \leq 1.5\,f_y/\sqrt{3} \quad 161\,\text{MPa} \leq 274.5\,\text{MPa}. \qquad (7.2)$$

For the fatigue assessment, the following conditions must be fulfilled:

Fig. 7.8 Measurement set-up for a fatigue test

Fig. 7.9 Mode of failure in a tested beam

$$\frac{\gamma_{Ff}\,\Delta\sigma_{E,2}}{\Delta\sigma_C/\gamma_{Mf}} \leq 1.0 \quad \frac{1.0\cdot 350}{400/1.0} \leq 1.0 \quad 0.875 \leq 1.0 \tag{7.3}$$

$$\frac{\gamma_{Ff}\,\Delta\tau_{E,2}}{\Delta\tau_C/\gamma_{Mf}} \leq 1.0 \quad \frac{1.0\cdot 161}{250/1.0} \leq 1.0 \quad 0.644 \leq 1.0 \tag{7.4}$$

It was verified for the combination of stress ranges $\Delta\sigma_{E,2}$ and $\Delta\tau_{E,2}$ that

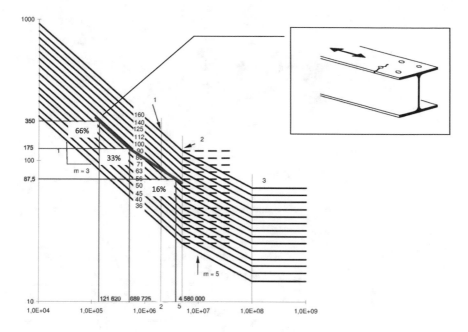

Fig. 7.10 Fatigue strength of the N1 beam for the individual stress ranges

$$\left(\frac{\gamma_{Ff}\,\Delta\sigma_{E,2}}{\Delta\sigma_C/\gamma_{Mf}}\right)^3 + \left(\frac{\gamma_{Ff}\,\Delta\tau_{E,2}}{\Delta\tau_C/\gamma_{Mf}}\right)^5 \leq 1.0 \quad 0.66992 + 0.11077 \leq 1.0 \quad 0.78069 \leq 1.0$$

(7.5)

Compliant with the STN EN 1993-1-9 standard, the curve obtained could be classified as the one for the group of details of a structural element with holes, subjected to bending and axial compression for a weakened perforated section; however, this classification is somewhat controversial as the beam is a composite of concrete and steel elements [6, 7].

7.5 Parametric Comparison of the Beams in the Fatigue Tests

The specimens with T-sections, either those with a smooth web edge or a different method of providing shear connection, manifested higher fatigue strengths in the tests than those with steel box sections. At the stress range of 20 ± 10 kN, the N1 beams could not withstand the expected number of 5 million stress-range cycles (Fig. 7.11).

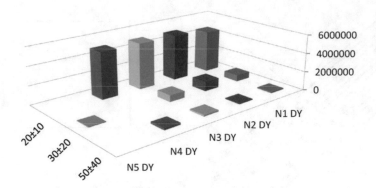

Fig. 7.11 Comparison of the N1–N5 beams in the fatigue tests

Table. 7.1 Parametric comparison of the fatigue test results on the N1–N5 beams

Beam No	Number of stress-range cycles	Stress range Δ(kN)
N1 DY	121 620	80
	689 725	40
	4 580 000	20
N2 DY	99 200	80
	1 020 000	40
	5 000 000	20
N3 DY	158 685	80
	708 000	40
	5 000 000	20
N4 DY	202 392	80
	–	40
	5 000 000	20
N5 DY	–	80
	699 920	40
	5 000 000	20

Table 7.1 presents the summary data on the fatigue strengths of the individual N1–N5 beams. Some of the fatigue tests are still ongoing; therefore, the table is incomplete. Fatigue failures most frequently occurred in zones with a notch, drilled hole or a weld.

References

1. P. Bocko, Posudzovanie životnosti nosných konštrukcií dopravných strojov a zariadení s využitím numerických a experimentálnych metód. Košice (2006)
2. J. Brooks, Pružnosť, zmrašťovanie, dotvarovanie a teplotné pohyby. Advanced Concrete Technology. Vlastnosti betónu, published by John Newman and Ban Seng Choo (2003). ISBN 0 7506 5104 0
3. H. Bryan, Fatigue in composite structures (Cambridge, 2003). ISBN 1 85573 608 – X.
4. M. Karmazínová, Experimental verification of fatigue resistance of steel expansion anchors to concrete loaded by cyclic tensile force with regard to particular failure modes. Conf. Exp. Appl. Mech. (Trans Tech Publications, Zurich, 2014). ISSN 1660-9336, ISBN 978-3-03835-028-6
5. M. Tomko, S. Kmeť, M. Bin, Experimental and theoretical behaviour analysis of steel suspension members subjected to tension and bending. Steel Compos. Struct. **13**(4), 343–365 (2012). ISSN 1229-9367
6. R. Ároch, I. Baláž, E. Chladný, S. Kmeť, J. Vičan, Navrhovanie oceľových konštrukcií podľa Eurokódov STN 1993-1-1:2006 a STN EN 1993-1-8:2007. Bratislava: Inžinierske konzultačné stredisko Slovenskej komory stavebných inžinierov, 212 (2007). ISBN 978-80-89113-35-4.
7. STN EN 1993-1 Eurokód 3: Navrhovanie oceľových konštrukcií, časť 9: Únava

Chapter 8
Modelling in the Abaqus Software Environment

8.1 Simulia Abaqus Software

The Simulia Abaqus product is a software package for the computational support of the design of a new product employing the finite-element analysis (FEA). Its first version was released in 1978 and both its name and logo are derived from ABAK, the ancient abacus calculation tool. In 2005 it became the cornerstone of a virtual laboratory called DSS SIMULIA. Hence, Abaqus is a set or suite of products that serve to wholly and realistically simulate a variety of physical disciplines, thereby making it possible to estimate and improve the qualities of a new product being designed, even before a prototype is built. Therefore, the costs of the developmental cycle can be dramatically reduced, and the functionality of various designs or even revolutionary concepts can be verified. This software tool can also be used to better explore the actual behaviour of structural members during an experiment. Stresses and deformations/strains can be observed at such places where observations or measurements were not feasible during the experiment.

Simulia Abaqus consists of the following modules:

Abaqus/CAE (Complete Abaqus Environment)

It is a complete software solution for the fast and effective modelling, pre-processing and analysing of finite elements. Next, it allows visualising the finite-element analysis results, their evaluating and post-processing. The environment for this pre/post-processor is based on a clearly-arranged and intuitive graphical interface, enabling its users to utilise all functions and advantages of the Abaqus package efficiently. One of the significant advantages of the Abaqus product is not only the possibility of importing the CAD geometry in all various formats but mainly its associative interface available for other programs to synchronise the CAD geometry and the computational model both ways, which incredibly saves the time necessary to complete one developmental/design loop during the design optimisation [1].

© The Author(s), under exclusive license to Springer Nature Switzerland AG 2021
V. Kvočák and D. Dubecký, *Research and Development of Deck Bridges*,
SpringerBriefs in Applied Sciences and Technology,
https://doi.org/10.1007/978-3-030-66925-6_8

Abaqus/Standard

Abaqus/Standard is a general-purpose finite-element analyser that employs solution technology ideal for static and low-speed dynamic events. It is used for all problems where the accuracy of numerical solutions is critically important. It can solve non-linear types of mechanical problems, and equally vibration- and noise-simulating computational problems.

Abaqus/Explicit

Abaqus/Explicit is a special-purpose finite-element analyser that serves mainly as a tool for simulating brief dynamic events, such as vehicle impact/crash testing, product drop testing, machining operations, ballistic impact or other kinds of penetration or breakdown testing. The explicit solver is well-suited for large-scale models with a vast number of elements and highly non-linear tasks, for instance, those with many contacts.

Abaqus/CFD (Computational Fluid Dynamics)

It is a powerful software tool intended for solving problems of computational fluid mechanics. Typical CFD tasks include, for example, seeking optimal solutions in the field of internal and external aerodynamics within the wide spectrum of Reynolds numbers or calculating the dynamic fluid effects on solid elements. Abaqus/CFD is also useful for thermo-mechanical simulations, including natural and forced convection heat transfer.

Because of the scalability of the software, each of the scientific areas above can be approached and solved separately; however, the Simulia Abaqus software package enables its users to communicate the results of any Abaqus analysis, collaborate and share the approved model structures.

In conclusion, Simulia Abaqus is an appropriate tool for modelling and analysing the non-linear behaviour of composite steel and concrete beams. Models of both materials can faithfully reflect their actual behaviours; for example, in crushing or cracking concrete, where the software can account for its plasticisation at the point of a crack. It makes it possible to monitor the stress–strain characteristics of the individual structural members and the points where cracks arise and sextend under cycling loading.

8.2   Preparing 3D Composite Beam Model

Modelling the task took place in the phase of *pre-processing*, which included creating an input file, geometry, editing material characteristics, meshing, setting the calculation parameters and boundary conditions. The geometry was modelled using the compatible CAD software. At the stage of pre-processing, points for the measurement of essential parameters were set, such as the acting forces, responses, stresses, deflections and others. For static problems, loading was simulated by moving loading

Fig. 8.1 A linear elastic
isotropic material

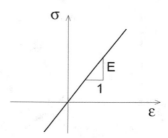

plates. Loading steps were selected at the levels of 20, 40, 60, 80, 100, 120, 140, and 160 kN. The preferred type of a finite element was a hexagonal prism—a Hexahedra element. Either the Newmark iteration method or the extended Hilber-Hughes-Taylor method can be used for the solution of the given non-linear problems. All calculations and evaluations of the results then happened in the *post-processing* phase.

8.3 Materials Used in Abaqus Model

8.3.1 Loading and Supporting Bearing Plates

Loading and supporting bearing plates represent the actual loading and supporting conditions of a structural element. As it is assumed that the amount of load is not as high as to cause plastic deformations in the plates, *SOLID 185*—a linear elastic isotropic material was used for modelling (Fig. 8.1), whose modulus of elasticity E, Poisson's ratio v and density σ can be edited [1].

8.3.2 Concrete Characteristics Used in Model

In the Abaqus software environment, concrete was defined and modelled as the *SOLID Concrete* group of materials with different characteristics, modified accordingly to suit the problems being solved. The »*concrete*« material that can correctly replicate the real behaviour of concrete in both tension and compression was selected. The software used simulates the initiation of cracks using the *CDP* (Concrete Damaged Plasticity) plastic model. Based on the experimentally determined $\sigma - \varepsilon$ curve for uniaxial tension and compression, it is possible to establish the correlation between cracking ($\tilde{\varepsilon}_t^{ck}$) due to uniaxial tension and crushing ($\tilde{\varepsilon}_t^{in}$) due to uniaxial compression. Figure 8.2 represents the values in the CDP model corresponding to tensile stresses ($\tilde{\varepsilon}_t^{ck}$) and those corresponding to compressive stresses ($\tilde{\varepsilon}_t^{in}$). A stress–strain diagram for concrete, which differs for the tension and compression regions, is

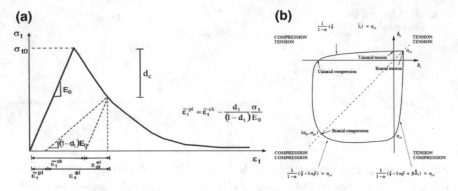

Fig. 8.2   a A $\sigma-\varepsilon$ tensile diagram for the CDP model **b** A stress–strain diagram for concrete

shown in Fig. 8.2. The diagram plotted from the material tests on the real specimens was considered in the computational FEA analysis.

8.3.3   Primary Rigid Steel Reinforcement and Secondary Reinforcement Bars

The primary rigid steel reinforcement and secondary reinforcement bars and stirrups were modelled as *Solid—8node185*. This model describes the stress–strain relationships of a reinforcing bar subjected to a uniaxial load. Two types of stress–strain diagrams are available: bilinear and multi-linear. The bilinear diagram describes the elastic and plastic behaviour of reinforcement. The plastic region can be defined by a rising branch with stiffening or by a constant branch representing perfect plasticity, as displayed in Fig. 8.3. The multi-linear diagram comprises four stages fully describing the stress–strain states in the reinforcement. The form of this diagram is fully symmetrical for both tension and compression.

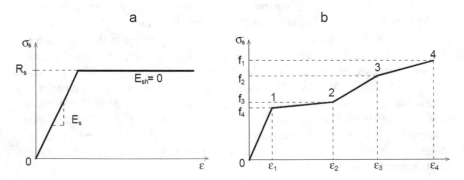

Fig. 8.3   a A bilinear stress–strain diagram **b** A multi-linear stress–strain diagram

The bilinear diagram with a rising branch was employed for the rigid steel reinforcement (Fig. 8.3).

For numerical simulations, the following material properties were used:

Primary rigid steel reinforcement

Nominal value of the yield strength of structural steel $f_y = 315, 3$ MPa
Modulus of elasticity of the reinforcing bars $\qquad E_s = 2800$ MPa
Modulus of elasticity of the structural steel $\qquad E_a = 210$ GPa

Secondary reinforcement bars

Characteristic value of the yield strength of reinforcing steel $f_{sk} = 500$ MPa
Modulus of elasticity of reinforcing steel $\qquad E_{sk} = 200$ GPa

Concrete

Nominal value of the cylinder compressive strength of concrete $f_c = 32, 17$ MPa
Modulus of elasticity of concrete $\qquad E_b = 33423, 8$ MPa
Nominal value of the flexural tensile strength of concrete $\qquad f_t = 3, 70$ MPa

8.4 Numerical Simulation Using the Abaqus Software

The numerical model was based on the experimental laboratory-tested specimens of beams. A three-dimensional view of a composite beam is presented in Fig. 8.4 below.

The composite beam was simulated in the Abaqus 6.11-2 software program. Advantage was taken of the symmetry of the beam and the applied load; therefore, only one half of the beam was modelled. The load was introduced through cylinders 70 mm in diameter, placed on a rigid 40 mm steel plate. The beam was then laid on line supports at both ends—the simulation of loading and supporting the beam corresponded to the same procedure during the laboratory experiment. 3D finite elements were used to form the real concrete and steel elements (of the filler-beam and reinforcement). The computational model contained approximately 180,000 elements and 41,000 nodes. Plastic deformations of the individual elements were taken into account.

(a)   **(b)**

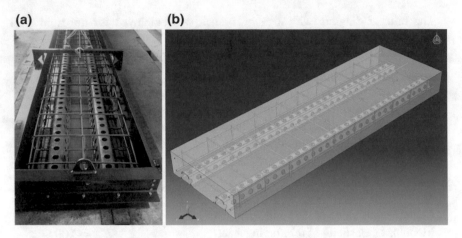

Fig. 8.4 a An experimental specimen **b** The Abaqus software model (half a beam)

8.5 The Composite Beam Model in Abaqus

In the Abaqus modelling, the *Concrete Damaged Plasticity* material model was
selected for the concrete, and the solution of *the Explicit dynamics* for the numerical
calculations, where the action time was 0.1 s. Contacts between the beam in the
support and the loading plate were regarded as a general Explicit contact. The contact
between the concrete and steel was considered to be rigid. The calculations lasted 42 h
on a two-processor computer, XEON 820—2 × 3, 0 GHz, 64 GB RAM processors,
where 12 system cores were used. At last, the simulation results were compared with
the experimental measurements.

The Abaqus software program simulates the initiation of cracks by a plastic
concrete model. In reality, there is no material at the point of a crack—the program,
though, works with a model where there is elastic-perfectly plastic material at the
point of a crack. The crack in the concrete manifests itself in the model of the beam
by the development of plastic deformations, as indicated in Fig. 8.5. The concrete
that transmitted compressive stresses is represented in Fig. 8.6.

At the point of the crack initiation towards the centre of the beam, where the
stresses were the highest, a plastic behaviour in the steel was consequently observed.
The stresses decreased towards the supports proportionally to the decrease in the load
on the beam, as can be seen in Fig. 8.7. Similarly, stresses arose in the reinforcement
bars, as is represented in Fig. 8.7.

Besides the relative deformations (strains) and stresses in a cross-section, deflections pose another significant indicator of structural behaviour. Their values depend
on the cross-sectional stiffness, and the most considerable deflection always occurs
in the middle of the beam span; however, its magnitude depends on the stiffness of
the whole structural member (Fig. 8.8).

The stiffness of the encased beams varied longitudinally: in those regions of the
beam where tensile stresses in the concrete did not exceed its tensile strength, the

Fig. 8.5 Development of plastic deformations in the concrete

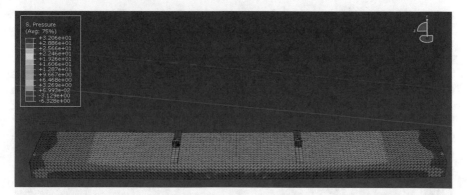

Fig. 8.6 Compressive stresses in the concrete

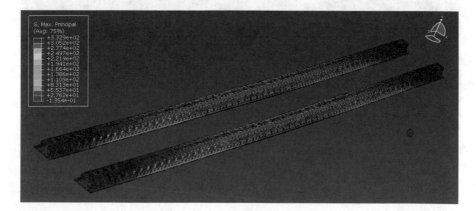

Fig. 8.7 A relative plastic behaviour in the rigid steel reinforcement

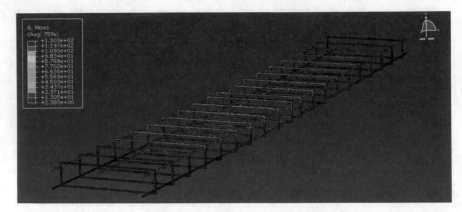

Fig. 8.8 Von Mises stresses in the reinforcement bars and stirrups

cross-sections exhibited the flexural stiffness EI_1, which was calculated with the consideration of the entire cross-section of the concrete segment. The cross-sections where cracks were fully developed exhibited the stiffness EI_2 and the tensile region of the concrete was regarded as excluded from the calculations and wholly neglected. In the regions between the two extremes, the flexural stiffness changed continuously. Deflections in the beam loaded by a force of 120 kN per cylinder are shown in Fig. 8.9.

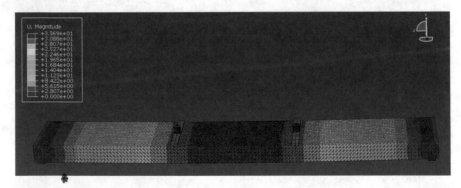

Fig. 8.9 Deflections in a symmetrically-loaded composite beam

8.6 Comparison of the Numerical Simulations with the Experimental Measurements

Deflections were monitored and recorded at several points of the beam during the experiments. The following graph compares the experimentally measured and numerically calculated mid-span deflections using the Abaqus software (Fig. 8.10). The difference between the two values for a 120kN load was 9%.

The stress–strain relationship in the steel sections dependent on the loading at a characteristic point is represented by the graph in Fig. 8.11. This point could be selected only where there is a node in the Abaqus model mesh. Depending on the size

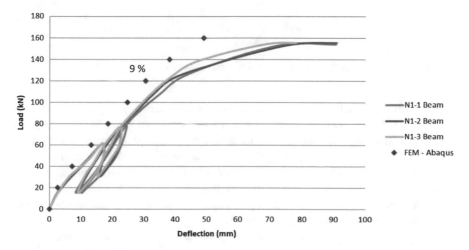

Fig. 8.10 Comparison of the experimentally measured and numerically calculated deflections

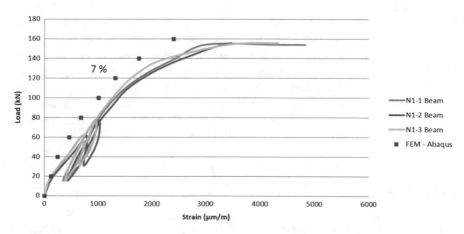

Fig. 8.11 Comparison of the stress–strain relationship in the steel section of a beam

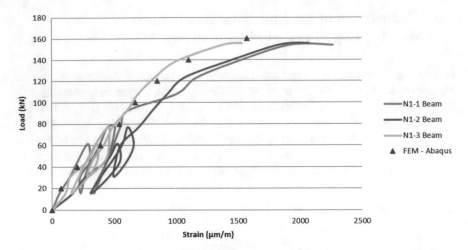

Fig. 8.12 Comparison of the stress–strain relationship in the concrete region of a beam

of the mesh, the spacing between the nodes was 15 mm. For this particular model, the modelling point was selected close to the measurement point in the laboratory tests. The difference between the values obtained in the experimental measurement and the numerical calculation using the Abaqus software for a 120kN load was 7%.

The stress–strain relationship in the concrete section dependent on the loading at a characteristic point is represented by graph in Fig. 8.12. The strains in the concrete measured during the experiment exhibited considerable differences due to the dissimilar opening of cracks in the concrete region.

Plastic behaviour of concrete at the point of cracks was assumed in the models presented. The initiation and propagation of cracks could be observed where plastic deformations occurred in the material. Nevertheless, it was not possible to monitor crack widths as the Abaqus modelling system does not allow this. Because of the vast number of elements and wide variability of shapes, it was necessary to use the explicit solver, which is suitable for large-scale models with many elements. Although the given models were not perfectly accurate, their comparison with the experimentally observed behaviour of the beam proved that they could be used to replicate the behaviour of the modelled constructions. A certain degree of inaccuracy contained in the model was caused by the usage of the material characteristics that had not been verified by experiment but determined according to the relevant provisions included in the STN EN 1992-1-1.

Reference

1. Simulia Abaqus Program Documentation, Part 21, Abaqus Analysis User's Manual. Materials available at: https://www.maths.cam.ac.uk/

Chapter 9
Outlook in the Field of Deck Bridges

The book in question pertains to some selected issues regarding composite steel and concrete structural members. It contains information on the state-of-the-art designs and constructional technologies in the field of composite structures in Slovakia and abroad. Particular attention in the book is drawn to deck bridges with encased filler-beams, with the aim of showing some new possibilities and challenges in such structures. This type of construction is well-suited to bridge obstacles over short spans, and it is currently exploited to bridge sluices, culverts, rivers, roads and other obstacles as wide as up to 18 m. Still, it is only the I-sections that are encased as the rigid reinforcement in the majority of cases. Another purpose of this work was to bring a new progressive modified and experimentally verified shape of steel section into the civil engineering world. The steel box section provides an improved method of composite action/shear connection, ensured by the holes that are made both in the top flange and the webs of the steel beam. Another advantage of the steel box section rests in its exceptional flexural stiffness, which can guarantee that the beam will bear the dead weight of the fresh concrete in the construction process if necessary.

Several specimens intended for the laboratory tests on composite steel and concrete beams were designed and introduced in the book. The book is a partial outcome of the extensive experimental research programme that has been done at the Institute of Structural Engineering at the Faculty of Civil Engineering at the Technical University of Košice, Slovakia. The research presented was undertaken on fifty experimental composite beams with rigid steel reinforcement, where five different types of composite action were tested. The book closely describes one specific type: the beam with a steel box section. Three methods of loading—static, long-term and fatigue—tests were carried out on the experimental beams. The testing procedure, along with the equipment used, was described in detail for each test. The experimental results were then compared with the analytical calculations according to the relevant technical standards currently in force. Applying the procedures and requirements compliant with the STN EN 1994-2 standard, the assumed values of bending resistance of the individual sections were calculated. The difference between the

V. Kvočák and D. Dubecký, *Research and Development of Deck Bridges*,
SpringerBriefs in Applied Sciences and Technology,
https://doi.org/10.1007/978-3-030-66925-6_9

theoretically calculated and experimentally measured deflections in the static test amounts to only 6.48%. This fact indicates that the new modified shapes of steel sections can be applied in constructions without the necessity to alter the design method stipulated by the STN EN 1994-1-1 standard.

Rheological changes in the composite steel and concrete beams were monitored in long-term tests. The tests revealed that the creep of concrete is most significant during the very first days after the load application. According to the experiment, 41% of the deflection occurred during the first week after the loading; therefore, it is always necessary to know the exact material characteristics at the time of loading the structural member.

Furthermore, the book outlines the test procedures and the results of the undertaken tests of material properties of the steel and concrete used in the experiments. The tensile strength of the steel was verified, as well as the compressive strength, flexural tensile strength, splitting tensile strength, and the static modulus of elasticity for the concrete in compression. Moreover, push-out tests were performed to verify the shear connection between the steel and concrete in the composite member, proving a sufficiently strong bond between the materials.

Following the actual behaviour of the composite beams, the experimental measurements and thorough analyses, a numerical model of the beam was created in the Abaqus software environment, faithfully simulating the behaviour of the beam under load. The difference between the numerically simulated and experimental deflection was 9% under a load of 120 kN. Material characteristics were carefully monitored during the experimental measurements and later employed in beam modelling. The Abaqus software can simulate various loading states in which material deformations can be monitored, even at the points that are usually unavailable during experiments. Thus, models created correctly with the aid of the software can help researchers and engineers in the future to simplify the creation and modification of a variety of other shapes of composite members without the necessity for experimental verifications.

Certain shortcomings of the steel box section were recognised during the research programme, which can be summarised in the following main points: the manufacturing of this section is very laborious, and it is considerably more substantial than the T-section, whereby increasing the production costs. Apart from that, there is the fatigue strength issue, which deserves serious attention. The steel box section is more susceptible to fatigue failure than the T-section. A crack initiated in the section at the point of the intermittent fillet weld, and then it propagated towards the hole in the side web. Fracture due to fatigue failure occurred in the bottom flange and the web where the hole was situated. However, these disadvantages were compensated for by a number of advantages that the given type of composite section can offer: the shear force the steel box section can transmit by far exceeds the shear connection, i.e. composite action other sections can provide. The steel box section, owing to its high stiffness, finds a whole range of applications in civil engineering constructions; for example, in bridges, concrete floor slabs with prefabricated profiled steel sheeting, flat roofs and many more. Today, when economic aspects are being discussed everywhere in the world, and Slovakia is not an exception, the selection of appropriate structural systems and building materials needs to be carefully considered so as to seek and achieve the optimum balance between their load-carrying capacity and price. When comparing and evaluating particular methods of ensuring the composite action between the steel and concrete sections in such structural members, it is clear that the application of modified steel sections can yield potential savings in the future. That said, future research should be encouraged in this field to examine, modify and optimise composite members.

Several conclusions and recommendations arising directly from the experimental research can be made, and they are presented below.

Recommendations for Further Research Direction

Based on the experience acquired, composite beams with steel box sections hold real potential for the practical application in constructions, although some efforts should be made to optimise their steel and concrete cross-sections. A big challenge will be the adjustment and possible lightening of a concrete deck that might lead to additional savings in material consumption. Some foreign theoretical research findings indicate that another stage in research investigations should be directed towards modern innovative VFT technologies with the utilisation of a variety of shear connection. Another direction in the research could be into permanent shuttering, where new progressive materials such as plastics, glass fibres or cement or magnesium-based materials could be utilised.

Moreover, numerical modelling is possible for such structures, allowing the simulation of fatigue and long-term tests. Such tests could be first run on the models and then compared with real experiments. Nonetheless, even though all these possible efforts would be extremely time-consuming, with correct results, they could bring considerable savings, and no more tests will be needed to verify further adjustments and modifications of the composite beams in the future.

The applications of concrete structures with encased steel filler-beams should extend to other fields, not only that of bridge-building. Steel box beams could well be used in floor slabs to lower headroom or in flat roofs. One line of research could also orientate towards the utilisation of thin-walled sections, taking advantage of the high tensile strength of steel but stabilised by concrete where such steel sections are encased.

Printed in the United States
By Bookmasters